FLOOD ASSESSMENT

Modeling and Parameterization

Innovations in Agricultural and Biological Engineering

FLOOD ASSESSMENT
Modeling and Parameterization

Edited by
Eric W. Harmsen, PhD
Megh R. Goyal, PhD, PE

Apple Academic Press Inc.
3333 Mistwell Crescent
Oakville, ON L6L 0A2 Canada

Apple Academic Press Inc.
9 Spinnaker Way
Waretown, NJ 08758 USA

© 2018 by Apple Academic Press, Inc.
First issued in paperback 2021
Exclusive worldwide distribution by CRC Press, a member of Taylor & Francis Group
No claim to original U.S. Government works

ISBN-13: 978-1-77463-048-8 (pbk)
ISBN-13: 978-1-77188-457-0 (hbk)

Library and Archives Canada Cataloguing in Publication

Flood assessment : modeling and parameterization / edited by Eric W. Harmsen, PhD, Megh R. Goyal, PhD, PE.

(Innovations in agricultural and biological engineering)
Includes bibliographical references and index.
Issued in print and electronic formats.

ISBN 978-1-77188-457-0 (hardcover).--ISBN 978-1-31536-592-3 (PDF)
1. Flood forecasting. 2. Floods--Risk assessment. 3. Hydrologic models. 4. Watersheds. I. Goyal, Megh Raj, editor II. Harmsen, Eric W., editor III. Series: Innovations in agricultural and biological engineering

GB1399.2.F66 2017   551.48'90112   C2017-902310-1   C2017-902311-X

Library of Congress Cataloging-in-Publication Data

Names: Harmsen, Eric W., editor. | Goyal, Megh Raj, editor.
Title: Flood assessment : modeling and parameterization / editors, Eric W. Harmsen, PhD, Megh R. Goyal, PhD, PE.
Description: Toronto ; New Jersey : Apple Academic Press, 2017. | Series: Innovations in agricultural & biological engineering | Includes bibliographical references and index.
Identifiers: LCCN 2017015046 (print) | LCCN 2017020899 (ebook) | ISBN 9781315365923 (ebook) | ISBN 9781771884570 (hardcover : alk. paper)
Subjects: LCSH: Flood forecasting. | Flood control. | Flood damage prevention. | Floods.
Classification: LCC GB1399.2 (ebook) | LCC GB1399.2 .F5355 2017 (print) | DDC 363.34/9372--dc23
LC record available at https://lccn.loc.gov/2017015046

Apple Academic Press also publishes its books in a variety of electronic formats. Some content that appears in print may not be available in electronic format. For information about Apple Academic Press products, visit our website at **www.appleacademicpress.com** and the CRC Press website at **www.crcpress.com**

CONTENTS

LIST OF CONTRIBUTORS

Megh R. Goyal, PhD, PE
Retired Faculty in Agricultural and Biomedical Engineering, University of Puerto Rico – Mayaguez Campus; and Senior Technical Editor-in-Chief in Agriculture Sciences and Biomedical Engineering, Apple Academic Press Inc., PO Box 86, Rincon – PR – 00677 – USA. E-mail: goyalmegh@gmail.com

Eric W. Harmsen, PhD
Professor, Department of Agricultural and Biosystems Engineering University of Puerto Rico – Mayaguez – Campus Mayaguez, PO Box 9000, Puerto Rico 00681-9000, USA; Mobile: +1 7879555102; Website: http://www.pragwater.com; E-mail: eric.harmsen@upr.edu; harmsen1000@gmail.com

Alejandra Rojas-González, PhD
School of Biosystems Engineering, Faculty of Engineering, Ciudad Universitaria Rodrigo Facio,P.O. BOX: 11501-2060, San Pedro de Montes de Oca, San José, Costa Rica; Phone: (506) 2511 5683, Cell phone: (506) 8887 7230; Email: alejandra.rojasgonzalez@ucr.ac.cr

Luz E. Torres-Molina, PhD
Assistant Professor, Department of Civil Engineering, University of Turabo in Puerto Rico. Mailing address: Reparto los Torres,1 Calle los Torres, Morovis, PR, 00687, USA. Mobile: +1 7873120409 Email: luz.torres2@upr.edu

LIST OF ABBREVIATIONS

AHPS	Advanced Hydrologic Prediction Service
AIC	akaike information criterion
AMS	annual maximum series
AR	autoregressive
ArcGIS	Arc Geographical Information Systems
ARIMA	autoregressive integrated moving average
ARMA	autoregressive moving average
CASA	collaborative adaptive sensing of the atmosphere
CASC2D	CASCade 2 dimensional sediment
CDA	contributing drainage area
COEM	Centro Residencial de Oportunidades Educativas de Mayagüez
CREST	Cooperative Remote Sensing Science and Technology Center
CRIM	Centro de Recaudación de Impuestos Municipales (Center for Municipal Tax Revenues of Puerto Rico)
CROEM	Centro Residencial de Oportunidades Educativas de Mayagüez
CRR	conceptual rainfall-runoff
DB	detection bias
DB	discrete bias
dBZ	decibels
dBZ	rain reflectivity expressed in logarithm scale
DEM	digital elevation map
DHSVM	distributed hydrology soils and vegetation model
DR	depth to rock
EPA	Environmental Protection Agency
ESRI	geographic information system company
ET	evapotranspiration
FAO	Food and Agriculture Organization

FAR	false alarm rate
FAS	flood alarm system
FEM	finite element method
FEMA	Federal Emergency Management Agency
FIS	Flood Insurance Study
GIS	geographical information system
GLUE	generalized likelihood uncertainty estimation
GOES	Geostationary Operational Environmental Satellite
GPD	generalized Pareto distribution
HAP	Hydrologic Rainfall Analysis Project
HE	hydro-estimator
HEC	Hydrologic Engineering Center
HEC-RAS	Hydrologic Engineering Center-River Analysis System
HMS	Hydrologic Modeling System
HR	hit rate
HRUs	hydrologic response units
IDW	inverse distance weighting
IE	infiltration excess
KWA	kinematic wave analogy
MA	moving average
MAE	mean absolute error
MAP	mean annual precipitation
MAR	mean annual rainfall
MatLab	Matrix Laboratory
MBDB	Mayagüez Bay Drainage Basin
MPE	multisensor precipitation estimation
MSE	mean square error
NAD	North American Datum
NCDC	National Climatic Data Center
NDFD	National Digital Forecast Database
NetCDF	Network Common Data Form
NEXRAD	Next Generation Radar
NLP	nonlinear optimization problem
NOAA	National Oceanic and Atmospheric Administration
NRCS	Natural Resources Conservation Service
NSF	National Science Foundation

NWCC	National Water and Climate Center
NWS	National Weather Service
OPPA	ordered physics-based parameter adjustment
OTG	off the grid
PBD	physically based distributed
PDF	probability distribution function
PDS	partial duration series
PET	potential or reference evapotranspiration
PI	performance index
POD	probability of detection
PPI	plan position indicator
PQPF	probabilistic quantitative precipitation forecast
PRAISE	prediction and rainfall amount inside storm events
PRMS	Precipitation Runoff Modeling System
PRWEB	Puerto Rico Water and Energy Balance
PRWRA	Puerto Rico Water Resources Authority
PRWRERI	Puerto Rico Water Resources and Environmental Research Institute
PRWRN	Puerto Rico Weather Radar Network
QP	quadratic programming
QPE	quantitative precipitation estimation
QPe-SUMS	quantitative precipitation estimation and segregation using multiple sensors
QPF	quantitative precipitation forecast
RETo	reference evapotranspiration or potential evapotranspiration
RHI	range height indication
RMSE	root mean squared error
$RMSE_{t+1}$	root mean squared error with a lead-time
RPS	ranked probability score
RQ	research question
RTFAS	Real Time Flood Alert System
SCAN	Soil Climate Analysis Network
SE	saturation excess
SHE	Systeme Hydrologique European
SQP	Sequence Quadratic Programming

SSURGO	Soil Survey Geographic Database
STD	standard deviation
TARS	Tropical Agriculture Research Station
TBSW	Testbed Subwatershed
TITAN	Thunderstorm Identification Tracking Analysis Nowcasting
TM	Thematic Mapper
TRMM	Tropical Rainfall Measuring Mission
UPRM	University of Puerto Rico Mayagüez Campus
USACE	U.S. Army Corps of Engineers
USDA	United Stated Department of Agriculture
USGS	U.S. Geological Survey
UTM	Universal Transverse Mercator
VPR	vertical profile reflectivity
WMO	World Meteorological Association
WSR-88D	Weather Surveillance Radar 1988 Doppler
W	West

LIST OF SYMBOLS

A	flow area
A	monthly regression coefficients
a, b	coefficients in the radar rain rate equation
Arc View	GIS software product
ASCII	data file in text format
B	monthly regression coefficients
BAG	Vflo filter
Bias	bias
BR	bias ratio
B_r	binary matrix
BR_{t+1}	bias ratio with a lead-time of $t + 1$
C	Celsius
C1	Rain Gauge station
cdf	cumulative distribution function
cms	cubic meters per second
d	distance
di	distance between centroids
E	expected value
e_a	actual vapor pressure
$e_{N,i}$	errors between rain gauge and NEXRAD
e_s	saturated vapor pressure
$e_{T,i}$	errors between rain gauge and TropiNet
ET_o	reference evapotranspiration
EW	exponential weighted
$F(\mathbf{x},\eta)$	Kriging function
$f(\mathbf{x})$	objective function
f_i	values of the scatter points
ft	feet
G	soil heat flux density
h	flow depth

$\hat{h}_{t+1,k(i,j)}$	rain estimated at time $(t+1)$
$\hbar_{t-1,k(i,j)}$	average reflectivity at time $(t-1)$, into the specific zone k
H	hit rate
h	hour
hr	hours
$h_{t-1,k(i,j)}$	reflectivity at time $(t-1)$, into the specific zone k
$h_{t-2,k(i,j)}$	reflectivity at time $(t-2)$, into the specific zone k
$h_{t,k(i,j)}$	reflectivity at time (t), into the specific zone k
I	infiltration
$I \times J$	possible combinations of forecast and observation
J	number of categories and therefore also the number of probabilities included in each forecast
K	hydraulic conductivity
k	zone into the forecast model
K_c	evapotranspiration crop coefficient
km	kilometer
kPa	kilopascal
K_s	saturated hydraulic conductivity
kw	kilo watts
La	average latitude
Lon	average longitude
m	meters
m	total number of columns of rainfall area
min	minutes
ML	maximum likelihood
mm	millimeters
MSE_N	mean square error between rain gauge and NEXRAD
MSE_T	mean square error between rain gauge and TropiNet
msl	mean sea level
n	Manning's hydraulic roughness
n	Manning's roughness factor
n	number of scatter points
n	variable number
n^*	total number of rows of rainfall area
N	North
N	total number of units

N0R	precipitation radar product (0)
N1P	precipitation radar product (1)
Ni	adjustment factor
N_i	NEXRAD data
N_{ii}	adjustment factor in the calibration
N_r	normalized reflectivity
O	model output with input parameters set at base values
O_1, O_2, O_3, O_J	observed values
o_j	cumulative probability of the observation in the i-th category or vector component
p	order of AR
P	value of input parameter
$p(y_t, O_j)$	joint distribution of forecast and observations
PR-1	CASA Student Testbed area 1
$p_x(x)$	Gaussian or normal distribution function
q	order of MA
Q	stream volumetric discharge
R	rain rate
R^2	Pearson correlation coefficient or coefficient of determination
R_a	extraterrestrial radiation
R_i	rain gauge data
R_n	net radiation
R_s	remote sensing
RXM-25	scientific name of TropiNet
s	slope
$S(d_{ij})$	model variogram evaluated at a distance to the distance between points i and j
s_f	Friction gradient
S_o	channel bed slope
S_r	relative sensitivity coefficient
SSE_N	sum squared error between rain gauge and NEXRAD
SSE_T	sum squared error between rain gauge and TropiNet
S22, S23, S24	CASA projects
T	temperature mean daily
t	time

T_{ave}	average air temperature
T_i	TropiNet data
T_{max}	maximum air temperature
T_{min}	minimum air temperature
TropiNet	Dual-polarization Doppler weather radar
u	channel velocity
U_0, U_1, U_2, U_N	regression coefficients
u_2	wind velocity at 2 m height
w_i	weight of the scatter points
$Vflo$	physically based hydrologic developed by Vieux and Associates, Inc
V	velocity
v	velocity between centroids
x	direction gradients
X	elevation
X, Y	generic variables on a probability plot
y	flow depth
y	prediction from the k-th simulation for Time, Peak and volume
Y	temperature
y_1, y_2, y_3, y_l	forecast values
y_j	cumulative probability assigned o the category or vector component
Y_m and O_m	cumulative forecast and observation
yr	year
$y_{t+l}(i,j)$	observed rainfall intensity at time t with lead-time
Z	reflectivity
Z_i	reflectivity in each pixel
Z_{max}	maximum reflectivity into the window
$Z_{max(t-1),k(i,j)}$	maximum reflectivity at time $(t-1)$
Z_{min}	reflectivity of 3 dBZ
$Z_{t-1,k(i,j)}$	ration between pixels with maximum reflectivity at time $(t-1)$
α	minimum value of reflectivity in the last two times
β	maximum value of reflectivity in the last two times
γ	psychometric constant in Penman Monteith equation

Δ	slope of the vapor pressure curve in Penman Monteith equation or change
$\delta1, \delta2, \delta3$	unknown parameters into the forecast model
Δx	columns of 9 pixels
Δy	rows of 9 pixels
θ	direction
Λ	Lagrange multiplicator
μ	mean
Q	scalars Controlling the infiltration, rainfall rate and hydraulic roughness
σ^2	variance
Φ	unknown variable into the forecast mod
$y_{t+1(i,j)}$	predicted rainfall intensity at time t with lead-time
ε_j	random error
$\varepsilon_{t,k(i,j)}$	random variable
(i,j)	Geographic position or coordinates latitude and longitude

PREFACE 1 BY ERIC W. HARMSEN

According to *The United Nations Office of Disaster Risk Reduction*, about 7000 people lose their lives and nearly 100 million people are adversely affected by floods each year worldwide. Flooding occurs in almost every part of the world and is the result of extreme rainfall. As an example, 30,000 people were killed from flooding in Venezuela in 1999. More than 12,000 people were killed in China during floods that occurred in 1980, 1996 and 1998; deaths from flooding exceeded 2600 people in Haiti during 2004; 2379 died in Bangladesh in 1988; 2311 died in Somalia in 1997; and 2001 died in India in 1994. Severe flooding caused economic losses totaling $122.5 billion in China, Korea, United States, Germany and Italy between 1991 and 2003. The most expensive flood on record occurred in China in 1998 costing $30 billion.

Flooding occurs when the runoff produced from heavy rainfall result in stream flows that exceed the flow capacity of the stream. Water overflows the stream banks covering areas of land that are normally dry. In 1940, Robert Horton described the process of rainfall infiltration and runoff from land surfaces. His conceptual model consisted of infiltration, which drops exponentially during a rainfall event until it reaches the long-term infiltration capacity of the soil. Given a constant rainfall rate, a point in time is reached, after the start of rainfall, when the infiltration capacity of the soil drops below the rain rate, and this is when surface runoff begins. Various factors affect the infiltration capacity of the soil including soil characteristics (texture, aggregation, bulk density, permeability, macropores, surface sealing, etc.), vegetation, antecedent moisture content, and other factors (e.g., land slope, air entrapment, surface roughness and temperature). Vegetation has a large influence on maintaining soil infiltration capacity. This is due in part to vegetation's ability to absorb rainfall energy that would otherwise pulverize surface aggregates, rapidly leading to surface sealing. The percent of impermeable area on the watershed also plays an important role in the amount of runoff and the peak flow rate that occurs near the watershed outlet.

A flash flood is defined as a flood that is associated with a weather event lasting 6 h or less. Flash floods are common in the Tropics, where rainfalls are associated with afternoon convective cloud development, often producing high intensity rainstorms that only last an hour or two. Other types of storms common to the Tropics are hurricanes, tropical waves and cold fronts originating in the upper latitudes. Rain cloud development can become amplified with the combination of sea breeze, convective, orographic and trade wind effects. In the coastal areas of tropical islands, it is not unusual to observe high intensity rainstorms. For example, at Camuy – Puerto Rico on November 24, 2015, a rainfall occurred that produced a storm total of 11.8 inches. For that location, the rainstorm corresponded with the 100-year return period (24-hour duration). Predicting extreme events like this one in time to issue a warning poses a particular challenge. The return period stated above was obtained from *National Oceanic and Atmospheric Administration (NOAA) Atlas 14 – Precipitation – Frequency Atlas of the United States*. Unfortunately during this century, documents such as this one will become less reliable as the frequency of extreme weather events increases due to global warming.

Flash flood warnings may call for an evacuation of an area, or provide guidance that a certain part of a city should be avoided, or that certain roads may be impassable. Flooding has been correlated with historic rainfall amounts. When a certain rainfall amount occurs, a warning can be issued, however, it may be too late to prevent loss of life and property. *Flood forecasting* attempts to predict flood levels at some time in the future (e.g., 1 to 2 h). The U.S. National Weather Service (NWS) in San Juan, Puerto Rico, uses the *Sacramento Soil Moisture Accounting* model along with a *quantitative precipitation forecast* (QPF) to evaluate flood potential and to guide decisions related to issuing flood warnings. The duration and intensity of the rain have an important influence on the peak stream or river flow. Unfortunately, the QPF does not provide reliable information related to duration and intensity of rainfall. Therefore, the NWS may evaluate several scenarios, for example, all the rain is assumed to fall in one hour, rainfall is equally distributed in a 3 h period, or rainfall randomly spread within a 3 h period. Using this approach it is possible to determine the rainfall distribution that produces the worst flooding.

To obtain high-resolution, site specific, event-specific flood information, a physically based numerical hydrologic model can be used. However, this type of model introduces other challenges and uncertainties. This book volume focuses on two detailed studies, which employed physically based hydrologic models. Despite the theoretical potential to obtain great accuracy with these models, much uncertainty may remain.

The Part I by Dr. Alejandra Rojas Gonzalez discusses flood prediction limitations in small watersheds with mountainous terrain and high rainfall variability. The hypothesis of the study is that it is possible to perform a small-scale, affordable model calibration, and then scale-up the parameters to a larger basin-scale model. Her study specifically addresses the following scientific questions: How is flow prediction affected by the spatial variability of point rainfall at scales below that of the typical resolution of radar-based products? How does parameter and hydrological model resolution affect the model's predictive capabilities and the errors of the hydrologic model? Would the assumptions developed for the small scale enhance the hydrologic predictability at larger scales?

Physically based hydrologic models can be given high-frequency input data and be run in near real-time. Unfortunately, there are occasions when real-time information does not provide enough time for the community to respond to a potentially dangerous situation. In this case a rainfall forecast must be made and continuously updated so that a flood prediction of one or two hours can be obtained. The study by Luz E. Torres Molina in Part II in this book volume describes the development of a stochastic model to forecast short-term rainfall for a tropical basin. The high-resolution rainfall data ($\approx$ 100-m) was derived using the TropiNet radar system at the University of Puerto Rico, Mayaguez Campus, representing possibly the only study of its kind in a tropical environment. The predicted short-term rainfall data was input into a hydrologic model land flood inundation levels were estimated at selected locations within the basin. Results of the rainfall and hydrologic forecasts are compared with observed data. The study also provides a prototype for a flood forecast alarm system.

It should be noted that the hydrologic model used in both studies described in the volume (Vflo) is limited to atmospheric, near-surface soil moisture, overland and stream flow processes, ignoring subsurface processes. Subsurface processes include aquifer recharge, groundwater flow and storage, and groundwater discharge to streams, lakes and the ocean. Groundwater discharge to streams is known as *stream base flow*. Since most hydrologic models do not include the subsurface component, stream base flow must be estimated and may contribute to stream flow uncertainty. A fully integrated surface/subsurface hydrologic model explicitly calculates the stream base flow from the ground water discharge component. Although, not included in this volume, a preliminary study using a fully integrated surface/subsurface hydrologic model has been conducted for the same basin considered in the two chapters in this book. Interested readers are encouraged to review the MS thesis of my former student Marcel Giovanni Prieto, *Development of a Regional Integrated Hydrologic Model for a Tropical Watershed* (M. G. Prieto, M.S. Thesis, 2007, Department of Civil Engineering and Surveying, University of Puerto Rico, Mayagüez Campus).

The purpose of this compendium is to contribute to a growing body of information about flood modeling in the Tropics. Some additional resources on the topic include:

- *Hydrologic Modeling of Land Processes in Puerto Rico Using Remotely Sensed Data* by J. F. Cruise and R. L. Miller
- *The Hydrology of the Humid Tropics* by E. Wohl et al.; *Physically Based Distributed Hydrologic Modeling of Tropical Catchments: Hypothesis Testing on Model Formation and Runoff Generation* by N. E. Abebe and F. L. Ogden
- *Rainfall-Runoff Modeling in a Flashy Tropical Watershed Using the Distributed HL-RDHM* model by A. Fares et al.
- *Development of a Regional Integrated Hydrologic Model for a Tropical Watershed* by Marcel Giovanni Prieto-Castellanos
- *Application of a Hydrological Model in a Data-Poor Tropical West African Catchment: A Case Study of the Densu Basin of Ghana* by E. O. Bekoe
- *Flooding Impacts and Modeling Challenges of Tropical Storms in Eastern Yemen* by K. Root and T. H. Papakos
- *Flood Prediction by Coupling KINEROS2 and HEC-RAS Models for Tropical Regions of Northern Vietnam* by H. Q. Nguyen, et al.

Other important books on the topic, not limited to tropical conditions include: *Hydrology and Flood Plain Analysis* by P. B. Bedient, W. C. Huber and B. E. Vieux

- *Distributed Hydrologic Modeling Using GIS* by B. E. Vieux and *Weather Radar Information and Distributed Hydrological Modeling* by Y. Tachikawa and B. E. Viuex.

I wish to extend my appreciation to the chapter authors who were advised by me during their PhD projects. Dr. Alejandra Rojas Gonzalez is currently an Assistant Professor in the Agricultural Engineering Department at the University of Costa Rica, and Dr. Luz Torres Molina is an Assistant Professor in the Department of Civil Engineering at the University of Turabo in Puerto Rico. I would also like to acknowledge the NOAA-CREST project (grant # NA11SEC4810004), which provided partial financial support for some of the research reported in the chapters and my participation on this book project. I especially would like to thank my colleague, Dr. Megh Raj Goyal, who assisted with the creation and editing of this book. Thanks also to the publishing staff at Apple Academic Press.

This book is dedicated to my dear, late brother Rick Harmsen, whose wise example guides me every day. The Bahá'í sacred Writings state: "*The progress of man's spirit in the divine world, after the severance of its connection with the body of dust, is through the bounty and grace of the Lord alone, or through the intercession and the sincere prayers of other human souls, or through the charities and important good works which are performed in its name.*" It is with this hope that I dedicate this work to my brother Rick.

The chapter authors and I are hopeful that this book volume will assist future researcher and practitioners in the field of flood modeling during the coming years, as they undoubtedly will face the challenge of increasing extreme weather events caused by a warming climate.

—Eric W. Harmsen, PhD

PREFACE 2 BY MEGH R. GOYAL

According to https: //en.wikipedia.org/wiki/Flood, a flood is *"an overflow of water that submerges land which is usually dry. The European Union (EU) Floods Directive defines a flood as a covering by water of land not normally covered by water. In the sense of 'flowing water', the word may also be applied to the inflow of the tide. Flooding may occur as an overflow of water from water bodies, such as a river or lake, in which the water overtops or breaks levees, resulting in some of that water escaping its usual boundaries, or it may occur due to an accumulation of rainwater on saturated ground in an areal flood. While the size of a lake or other body of water will vary with seasonal changes in precipitation and snow melt, these changes in size are unlikely to be considered significant unless they flood property or drown domestic animals. Floods can also occur in rivers when the flow rate exceeds the capacity of the river channel, particularly at bends or meanders in the waterway. Floods often cause damage to homes and businesses if they are in the natural flood plains of rivers. While riverine flood damage can be eliminated by moving away from rivers and other bodies of water, people have traditionally lived and worked by rivers because the land is usually flat and fertile and because rivers provide easy travel and access to commerce and industry. Some floods develop slowly, while others such as flash floods, can develop in just a few minutes and without visible signs of rain. Additionally, floods can be local, impacting a neighborhood or community, or very large, affecting entire river basins."*

In general, we do not like floods because of their negative impacts on our daily life. *"Ferdinand Quinones and Karl G. Johnson, 1987. The Floods of May 17–18, 1985 and October 6–7, 1985 in Puerto Rico. US Geological Survey Open-file Report 87–123. U.S. Geological Survey Books and Open-File Reports Federal Center"* indicates that *"During 1985, severe floods occurred twice throughout Puerto Rico resulting in significant losses in life and property. The first event occurred during May 15–19, when a low-pressure system resulted in precipitation totals*

*exceeding 14 inches throughout most of south-central and eastern Puerto
Rico. A second event was produced by a tropical depression that affected
south-central Puerto Rico during October 6–7. Landslides and collapses
of several key bridges during the October floods resulted in the death of
as many as 170 people. Property losses from both floods were estimated at
about 162 million dollars. A nearly stationary tropical depression affected
Puerto Rico during October 6–7, 1985, resulting in 24-hour precipitation
totals of as much as 23 inches and severe floods along the south-central
coastal areas. The floods of October 6–7, 1985, affected mostly rural areas
in southern Puerto Rico, but caused significant loss of life and widespread
property damages. Landslides near Ponce, the collapse of a bridge at Rio
Coamo, and the destruction of homes near Ponce resulted in about 170
fatalities and more than 125 million dollars in damages. Flooding was
also severe at Barceloneta on the north coast. Recurrence intervals equal
to or greater than 100 years were estimated for peak discharges at several
index stations".*

I am an eyewitness of the second flood on October 6–7 of 1985 in
Ponce. The flood level was almost 2.5 feet inside our home, and we were
rescued to a higher elevated area. We lost almost all our property worth
$20,000, and my family was shocked. I like to share with the readers the
thoughts from "Gleanings from the Writings of Baha'u'llah, Bahá'í Pub,
2005": *"For every one of you his paramount duty is to choose for himself
that on which no other may infringe and none usurp from him. Such a thing
– and to this the Almighty is my witness – is the love of God, could ye but
perceive it. Build ye for yourselves such houses as the rain and floods can
never destroy, which shall protect you from the changes and chances of
this life. This is the instruction of Him Whom the world hath wronged and
forsaken."* Was my home not properly built?

The mission of this book volume is to serve as a reference manual for
graduate and undergraduate students of agricultural, biological and civil
engineering; as well as those in horticulture, soil science, crop science
and agronomy. I hope that it will be a valuable reference for professionals
who work with flood management; and for professional training institutes,
technical agricultural centers, irrigation centers, Agricultural Extension
Service, and other agencies.

At the 49th annual meeting of the Indian Society of Agricultural staff Engineers at Punjab Agricultural University during February 22–25 of 2015, a group of ABEs convinced me that there is a dire need to publish book volumes on focus areas of agricultural and biological engineering (ABE). This is how the idea was born for a new book series titled *"Innovations in Agricultural & Biological Engineering."*

My longtime colleague, Dr. Eric W. Harmsen, joins me as a Lead Editor of this volume. Dr. Harmsen holds exceptional professional qualities with his expertise in agricultural hydrology during the last 35 years, in addition his role as research scientist at the University of Puerto Rico – Mayaguez Campus. His generous offer and contributions by his students, Alejandra Rojas-Gonzalez and Luz E. Torres-Molina, to the contents and quality of this book have been invaluable.

Abdu'l-Baha in the book The Chosen Highway, Lady Blomfield, George Ronald Pub Ltd (2007)" righty describes our cooperation in His holy words as *"those who work singly are as drops, but, when united, they will become a vast river carrying the cleansing water of life into the barren desert places of the world. Before the power of its rushing flood, neither misery, nor sorrow, nor any grief will be able to stand. Be united!"*

We would like to thank editorial staff, Sandy Jones Sickels, Vice President, and Ashish Kumar, Publisher and President at Apple Academic Press, Inc., for making every effort to publish the book when the diminishing water resources are a major issue worldwide. Special thanks are due to the AAP Production Staff for typesetting.

We request that readers offer us your constructive suggestions that may help to improve the next edition. The reader can order a copy of this book for the library, the institute or for a gift from "http://appleacademicpress. com."

Our Almighty God, owner of natural resources, must be very happy on publication of this book. As an educator, there is a piece of advice to one and all in the world: *"Permit that our almighty God, our Creator and excellent Teacher, help us to solve and manage problems in flood management with His Grace."*

—Megh R. Goyal, PhD, PE

WARNING/DISCLAIMER

PLEASE READ CAREFULLY

The goal of this book volume on *Flood Assessment: Modeling and Parameterization* is to present challenges, issues and new technologies. The editors, the contributing authors, the publisher and the printer have made every effort to make this book as accurate as possible. However, there still may be grammatical errors or mistakes in the content or typography. Therefore, the contents in this book should be considered as a general guide and not a complete solution to address any specific situation.

The editors, the contributing authors, the publisher and the printer shall have neither liability nor responsibility to any person, any organization or entity with respect to any loss or damage caused, or alleged to have caused, directly or indirectly, by information or advice contained in this book. Therefore, the purchaser/reader must assume full responsibility for the use of the book or the information therein.

The mention of commercial brands and trade names are only for technical purposes. It does not mean that a particular product is endorsed over to another product or equipment not mentioned. Author, cooperating authors, educational institutions, and the publisher Apple Academic Press Inc. do not have any preference for a particular product.

All weblinks that are mentioned in this book were active on December 31, 2016. The editors, the contributing authors, the publisher and the printing company shall have neither liability nor responsibility, if any of the weblink is inactive at the time of reading of this book.

ABOUT LEAD EDITOR

 **Eric W. Harmsen, PhD**
Professor, Department of Agricultural and Biosystems Engineering, University of Puerto Rico – Mayaguez – Campus Mayaguez, Puerto Rico

Dr. Eric W. Harmsen obtained his BS and MS degrees in Agricultural Engineering from Michigan State University, PhD degree from University of Wisconsin-Madison, and performed a Post-doctoral study at North Carolina State University. Currently, he is a Professor in the Department of Agricultural and Biosystems Engineering, University of Puerto Rico, Mayaguez Campus. He teaches courses in agricultural hydrology, agroclimatology, and irrigation. His professional interests include measurement and modeling all components of the hydrologic cycle; remote sensing of water and energy budgets in the tropics; and hydrology, irrigation, and agroclimatology. Dr. Harmsen maintains a website, which provides daily operational water and energy balance components as well as soil and water parameters related to drought and crop health in Puerto Rico (http://www.pragwater.com).

Dr. Harmsen's publications cover a wide range of topics, including numerical simulation and field measurement of rainfall, evapotranspiration, surface runoff, aquifer recharge, soil moisture, weather-related variables, and groundwater and vadose zone processes. He co-edited the book Evapotranspiration: Principles and Applications for Water Management, has published 5 book chapters, 28 peer-reviewed journal articles, and 41 conference proceedings. Readers may contact him at: eric.harmsen@upr.edu, harmsen1000@gmail.com.

ABOUT CO-EDITOR

Megh R Goyal, PhD, PE
*Retired Professor in Agricultural and Biomedical
Engineering, University of Puerto Rico,
Mayaguez Campus Senior Acquisitions Editor,
Biomedical Engineering and Agricultural Science,
Apple Academic Press, Inc.*

Megh R. Goyal, PhD, PE, is a Retired Professor in Agricultural and Biomedical Engineering from the General Engineering Department in the College of Engineering at University of Puerto Rico–Mayaguez Campus; and Senior Acquisitions Editor and Senior Technical Editor-in-Chief in Agriculture and Biomedical Engineering for Apple Academic Press Inc.

He has worked as a Soil Conservation Inspector and as a Research Assistant at Haryana Agricultural University and Ohio State University. He was the first agricultural engineer to receive the professional license in Agricultural Engineering in 1986 from the College of Engineers and Surveyors of Puerto Rico. On September 16, 2005, he was proclaimed as "Father of Irrigation Engineering in Puerto Rico for the twentieth century" by the ASABE, Puerto Rico Section, for his pioneering work on micro irrigation, evapotranspiration, agroclimatology, and soil and water engineering. During his professional career of 45 years, he has received many prestigious awards. A prolific author and editor, he has written more than 200 journal articles and textbooks and has edited over 50 books. He received his BSc degree in engineering from Punjab Agricultural University, Ludhiana, India; his MSc and PhD degrees from Ohio State University, Columbus; and his Master of Divinity degree from Puerto Rico Evangelical Seminary, Hato Rey, Puerto Rico, USA. Readers may contact him at: goyalmegh@gmail.com.

ABOUT THE AUTHORS

Luz E. Torres Molina, PhD
*Assistant Professor, Dept. of Civil Engineering,
School of Engineering, Universidad del Turabo,
Gurabo, Puerto Rico*

Dr. Torres-Molina has a BS degree in Civil Engineering from the Universidad Industrial de Santander (Colombia). Between 2002 and 2004 she obtained an MS degree in Environmental and Water Resources (University of Puerto Rico),working with flood waters in compound channels. In 2005, she began working as a consulting engineer in water resources: water distribution lines, water resources design and technical analysis, several designs for flood control, river regulation, erosion and sediment control analysis, water/sewer system analysis, ground water, drainage, and scour analysis, with CA Engineering and CMA Architects & Engineers. In 2014 she completed her PhD studies in the Dept. of Civil and Environmental Engineeringat the University of Puerto Rico-Mayaguez Campus. Since 2015 she has been working as anassistant professor in the Department of Civil Engineering at the Universidad del Turabo in Puerto Rico.

Alejandra M. Rojas González, PhD
*Associate Professor, Department of Biosystems
Engineering, University of Costa Rica, Rodrigo
Facio Campus, San José, Costa Rica*

Dr. Alejandra M. Rojas obtained her BS degree in Agricultural Engineering from the University of Costa Rica, and her Master and PhD degrees in Civil Engineering from the University of Puerto Rico-Mayaguez. Currently, she is Associate Professor in the Department of Biosystems Engineering at the University of Costa Rica, Rodrigo Facio Campus. She teaches courses in applied hydrology, geographical information systems, and remote sensing applications. Her professional interests include measurement and modeling of all components of hydrological cycle, remote sensing of water, integrated water resources management, hydraulics modeling, and flood risk reduction.

Dr. Rojas has conducted research in flood risk reduction in areas with high rainfall intensities, and shehas been developing research with Costa Rican institutions to improve the management of water resources and water quality in systems impacted by droughts.

OTHER BOOKS ON AGRICULTURAL AND BIOLOGICAL ENGINEERING BY APPLE ACADEMIC PRESS, INC.

Management of Drip/Trickle or Micro Irrigation
Megh R. Goyal, PhD, PE, Senior Editor-in-Chief

Evapotranspiration: Principles and Applications for Water Management
Megh R. Goyal, PhD, PE, and Eric W. Harmsen, Editors

Book Series: Research Advances in Sustainable Micro Irrigation
Senior Editor-in-Chief: Megh R. Goyal, PhD, PE

Book Series: Innovations and Challenges in Micro Irrigation
Senior Editor-in-Chief: Megh R. Goyal, PhD, PE

Volume 1: Principles and Management of Clogging in Micro Irrigation
Volume 2: Sustainable Micro Irrigation Design Systems for Agricultural
Crops: Methods and Practices
Volume 3: Performance Evaluation of Micro Irrigation Management:
Principles and Practices
Volume 4: Potential Use of Solar Energy and Emerging Technologies in
Micro Irrigation
Volume 5: Micro Irrigation Management: Technological Advances and
Their Applications
Volume 6: Micro Irrigation Engineering for Horticultural Crops: Policy
Options, Scheduling, and Design
Volume 7: Micro Irrigation Scheduling and Practices
Volume 8: Engineering Interventions in Sustainable Trickle Irrigation:
Water Requirements, Uniformity, Fertigation, and Crop
Performance

Book Series: Innovations in Agricultural and Biological Engineering
Senior Editor-in-Chief: Megh R. Goyal, PhD, PE
• Dairy Engineering: Advanced Technologies and their Applications
• Developing Technologies in Food Science: Status, Applications, and
Challenges
• Emerging Technologies in Agricultural Engineering
• Engineering Interventions in Agricultural Processing
• Engineering Practices for Agricultural Production and Water
Conservation: An Interdisciplinary Approach
• Flood Assessment: Modeling and Parameterization
• Food Engineering: Modeling, Emerging Issues and Applications.
• Food Process Engineering: Emerging Trends in Research and Their
Applications
• Food Technology: Applied Research and Production Techniques
• Modeling Methods and Practices in Soil and Water Engineering
• Processing Technologies for Milk and Milk Products: Methods,
Applications, and Energy Usage
• Soil and Water Engineering: Principles and Applications of Modeling
• Soil Salinity Management in Agriculture: Technological Advances
and Applications

- Technological Interventions in Management of Irrigated Agriculture
- Technological Interventions in the Processing of Fruits and Vegetables
- State-of-the-Art Technologies in Food Science
- Sustainable Biological Systems for Agriculture
- Novel Dairy Processing Technologies: Techniques, Management, and Energy Conservation
- Technological Interventions in Dairy Science: Innovative Approaches in Processing, Preservation, and Analysis of Milk Products
- Engineering Interventions in Foods and Plants

EDITORIAL

Apple Academic Press, Inc., (AAP) is publishing book volumes in the specialty areas as part of *Innovations in Agricultural and Biological Engineering book series*, over a span of 8 to 10 years. These specialty areas have been defined by *American Society of Agricultural and Biological Engineers* (http://asabe.org).

The mission of this series is to provide knowledge and techniques for Agricultural and Biological Engineers (ABEs). The series aims to offer high-quality reference and academic content in Agricultural and Biological Engineering (ABE) that is accessible to academicians, researchers, scientists, university faculty, and university-level students and professionals around the world. The following material has been edited/modified and reproduced below *"Goyal, Megh R., 2006. Agricultural and biomedical engineering: Scope and opportunities. Paper Edu_47 at the Fourth LACCEI International Latin American and Caribbean Conference for Engineering and Technology (LACCEI' 2006): Breaking Frontiers and Barriers in Engineering: Education and Research by LACCEI University of Puerto Rico – Mayaguez Campus, Mayaguez, Puerto Rico, June 21–23."*

WHAT IS AGRICULTURAL AND BIOLOGICAL ENGINEERING (ABE)?

"Agricultural Engineering (AE) involves application of engineering to production, processing, preservation and handling of food, fiber, and shelter. It also includes transfer of technology for the development and welfare of rural communities," according to http://isae.in." *ABE is the discipline of engineering that applies engineering principles and the fundamental concepts of biology to agricultural and biological systems and tools, for the safe, efficient and environmentally sensitive production, processing, and management of agricultural, biological, food, and natural resources systems,"* according to http://asabe.org. *"AE is the branch of engineering*

involved with the design of farm machinery, with soil management, land development, and mechanization and automation of livestock farming, and with the efficient planting, harvesting, storage, and processing of farm commodities," definition by: http://dictionary.reference.com/browse/agricultural+engineering.

"*AE incorporates many science disciplines and technology practices to the efficient production and processing of food, feed, fiber and fuels. It involves disciplines like mechanical engineering (agricultural machinery and automated machine systems), soil science (crop nutrient and fertilization, etc.), environmental sciences (drainage and irrigation), plant biology (seeding and plant growth management), animal science (farm animals and housing) etc.,*" by: http://www.ABE.ncsu.edu/academic/agricultural-engineering.php.

"According to https://en.wikipedia.org/wiki/Biological_engineering: "*BE (Biological engineering) is a science-based discipline that applies concepts and methods of biology to solve real-world problems related to the life sciences or the application thereof. In this context, while traditional engineering applies physical and mathematical sciences to analyze, design and manufacture inanimate tools, structures and processes, biological engineering uses biology to study and advance applications of living systems.*"

SPECIALTY AREAS OF ABE

Agricultural and Biological Engineers (ABEs) ensure that the world has the necessities of life including safe and plentiful food, clean air and water, renewable fuel and energy, safe working conditions, and a healthy environment by employing knowledge and expertise of sciences, both pure and applied, and engineering principles. Biological engineering applies engineering practices to problems and opportunities presented by living things and the natural environment in agriculture. BA engineers understand the interrelationships between technology and living systems, have available a wide variety of employment options. "*ABE embraces a variety of following specialty areas,*" http://asabe.org. As new technology and information emerge, specialty areas are created, and many overlap with one or more other areas.

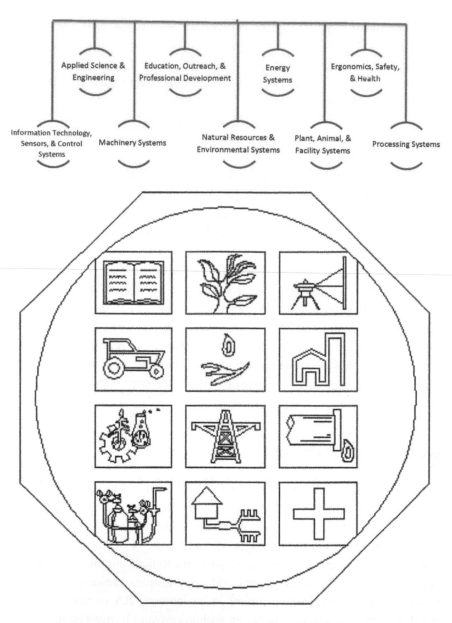

1. **Aquacultural Engineering**: ABEs help design farm systems for raising fish and shellfish, as well as ornamental and bait fish. They specialize in water quality, biotechnology, machinery, natural resources, feeding and ventilation systems, and sanitation. They

seek ways to reduce pollution from aquacultural discharges, to reduce excess water use, and to improve farm systems. They also work with aquatic animal harvesting, sorting, and processing.

2. **Biological Engineering** applies engineering practices to problems and opportunities presented by living things and the natural environment.

3. **Energy:** ABEs identify and develop viable energy sources – biomass, methane, and vegetable oil, to name a few – and to make these and other systems cleaner and more efficient. These specialists also develop energy conservation strategies to reduce costs and protect the environment, and they design traditional and alternative energy systems to meet the needs of agricultural operations.

4. **Farm Machinery and Power Engineering**: ABEs in this specialty focus on designing advanced equipment, making it more efficient and less demanding of our natural resources. They develop equipment for food processing, highly precise crop spraying, agricultural commodity and waste transport, and turf and landscape maintenance, as well as equipment for such specialized tasks as removing seaweed from beaches. This is in addition to the tractors, tillage equipment, irrigation equipment, and harvest equipment that have done so much to reduce the drudgery of farming.

5. **Food and Process Engineering:** Food and process engineers combine design expertise with manufacturing methods to develop economical and responsible processing solutions for industry. Also food and process engineers look for ways to reduce waste by devising alternatives for treatment, disposal and utilization.

6. **Forest Engineering**: ABEs apply engineering to solve natural resource and environment problems in forest production systems and related manufacturing industries. Engineering skills and expertise are needed to address problems related to equipment design and manufacturing, forest access systems design and construction; machine-soil interaction and erosion control; forest operations analysis and improvement; decision modeling; and wood product design and manufacturing.

7. **Information and Electrical Technologies Engineering** is one of the most versatile areas of the ABE specialty areas, because it is

applied to virtually all the others, from machinery design to soil testing to food quality and safety control. Geographic information systems, global positioning systems, machine instrumentation and controls, electromagnetics, bioinformatics, biorobotics, machine vision, sensors, spectroscopy: These are some of the exciting information and electrical technologies being used today and being developed for the future.

8. **Natural Resources:** ABEs with environmental expertise work to better understand the complex mechanics of these resources, so that they can be used efficiently and without degradation. ABEs determine crop water requirements and design irrigation systems. They are experts in agricultural hydrology principles, such as controlling drainage, and they implement ways to control soil erosion and study the environmental effects of sediment on stream quality. Natural resources engineers design, build, operate and maintain water control structures for reservoirs, floodways and channels. They also work on water treatment systems, wetlands protection, and other water issues.

9. **Nursery and Greenhouse Engineering**: In many ways, nursery and greenhouse operations are microcosms of large-scale production agriculture, with many similar needs – irrigation, mechanization, disease and pest control, and nutrient application. However, other engineering needs also present themselves in nursery and greenhouse operations: equipment for transplantation; control systems for temperature, humidity, and ventilation; and plant biology issues, such as hydroponics, tissue culture, and seedling propagation methods. And sometimes the challenges are extraterrestrial: ABEs at NASA are designing greenhouse systems to support a manned expedition to Mars!

10. **Safety and Health:** ABEs analyze health and injury data, the use and possible misuse of machines, and equipment compliance with standards and regulation. They constantly look for ways in which the safety of equipment, materials and agricultural practices can be improved and for ways in which safety and health issues can be communicated to the public.

11. **Structures and Environment:** ABEs with expertise in structures and environment design animal housing, storage structures, and greenhouses, with ventilation systems, temperature and humidity controls, and structural strength appropriate for their climate and purpose. They also devise better practices and systems for storing, recovering, reusing, and transporting waste products.

CAREER IN AGRICULTURAL AND BIOLOGICAL ENGINEERING

One will find that university ABE programs have many names, such as biological systems engineering, bioresource engineering, environmental engineering, forest engineering, or food and process engineering. Whatever the title, the typical curriculum begins with courses in writing, social sciences, and economics, along with mathematics (calculus and statistics), chemistry, physics, and biology. Student gains a fundamental knowledge of the life sciences and how biological systems interact with their environment. One also takes engineering courses, such as thermodynamics, mechanics, instrumentation and controls, electronics and electrical circuits, and engineering design. Then student adds courses related to particular interests, perhaps including mechanization, soil and water resource management, food and process engineering, industrial microbiology, biological engineering or pest management. As seniors, engineering students team up to design, build, and test new processes or products.

For more information on this series, readers may contact:

Ashish Kumar, Publisher and President
Sandy Sickels, Vice President
Apple Academic Press, Inc.
Fax: 866-222-9549
E-mail: ashish@appleacademicpress.com
http://www.appleacademicpress.com/
publishwithus.php

Megh R. Goyal, PhD, PE
Book Series Senior
Editor-in-Chief
Innovations in Agricultural and Biological Engineering
E-mail: goyalmegh@gmail.com

PART I

FLOOD PREDICTION LIMITATIONS IN SMALL WATERSHEDS

CHAPTER 1

FLOOD PREDICTION LIMITATIONS IN SMALL WATERSHEDS: INTRODUCTION[1,2]

ALEJANDRA M. ROJAS-GONZÁLEZ

CONTENTS

1.1 INTRODUCTION

Due to the complex terrain and the tropical climate influence, Puerto Rico is characterized by small watersheds, high rainfall intensity and spatial variability. The rainfall anomalies are produced by tropical waves, low pressure depressions, tropical storms, and hurricanes capable of producing flash flood in susceptible areas. As part of the model configuration, rainfall

[1] This chapter is an edited version from *Alejandra María Rojas González, 2012. Flood prediction limitations in small watersheds with mountainous terrain and high rainfall variability. Unpublished PhD Thesis for Department of Civil Engineering and Surveying, University of Puerto Rico – Mayagüez Campus.*

[2] Numbers in brackets refer to the references at the end of this book.

must be distributed over the model domain. Different theoretical methods are available to spatially distribute rainfall over a watershed. However, there is not typically enough rain gauge density to calculate the associated bias, and to obtain spatial variability of point rainfall at scales below the typical resolution of the radar-based products (2×2 km^2), archived with the Next Generation Radar (NEXRAD) level 3.

New emerging radar technologies are being developed by the Student Test Bed of the Center for Collaborative Adaptive Sensing of the Atmosphere [22] in Puerto Rico and will be available for flash flood predictions. This new radar technology promises to revolutionize the way rainfall is detected, monitored and predicted, creating a dense sensor network of low-powered radars that overcome curvature blockage and significantly enhance resolution. This network will monitor the lower atmosphere where the principal atmospheric phenomena occur. The first step in the technology development has been the PR-1 radar located at the roof top of the Stefani building at University of Puerto Rico, Mayagüez Campus. The PR-1 radar is marine radar adapted to sense reflectivity with an average pixel size of 150 m and the maximum coverage range of 25 km.

An important step for the hydrologic community and Puerto Rico in general will be the use of these advanced technologies as input to real-time flash flood prediction systems. Real-time flash flood estimates can allow decision makers to implement emergency plans only when it is necessary, since unnecessary preparations and evacuations are very costly. The technique also allows decision makers to better focus the emergency measures due to variable rainfall patterns. Since in the tropical region the locations, where flood waters concentrate, tend to vary in time and space. Rain gauge density is generally not sufficient to capture spatial variability at the NEXRAD radar subpixel scale and the new radar technology will help to fill gaps between rain gauges. Some methods for removing the systematic bias between radar and rain gauges are applied today. However, it is not known how much the intrinsic error due to spatial variability at the radar subpixel scale limits the reliability of the data for use in hydrologic models. Some scientific questions arise where complex terrain and climatological conditions increase the spatially dependent bias.

How does rainfall spatial distribution affect the hydrologic response in small sub watersheds? How can adjustments be made to radar rainfall estimates when there are not sufficient numbers of rain gauges within the network? Under these conditions, how can we produce reliable hydrologic estimates in small areas where high spatial variability exists? These questions are essential when using fully physics-based distributed hydrologic models, because the goal of their use is to produce accurate flood predictions at any location upstream of the watershed outlet.

Few studies have been conducted in Puerto Rico to forecast real-time rainfall and runoff. In 1996, the US Geological Survey (USGS) developed a real time rainfall runoff simulation for Carraízo reservoir basin allowing the estimation of water volumes at the reservoir from the rainfall and discharge data that is being obtained from the network stations inside the basin [94].

The National Weather Service (NWS) establishes Flash Flood Guidance estimates in real time based on the Sacramento soil moisture accounting model. Flash Flood Guidance is performed by region or River Forecast Center, and Puerto Rico belongs to the South-east River Forecast Center. The analysis allows for the development of the curves that relate threshold runoff to flash flooding of a given duration as a function of soil moisture deficit [36, 81, 97, 101]. Vieux and Vieux [135] tested a physics-based distributed model in the Loíza basin of Puerto Rico. A long-term and event-based simulation was conducted to calibrate the streamflow volume. The soil moisture values calculated in the long-term model were fed back into the event-based simulation to enhance the calibration for several individual storm events. A sensitivity analysis to initial soil moisture showed some persistence in antecedent soil conditions, with about one year of warm up the model to obtain stable results.

To establish a flood alarm system in Puerto Rico, first it is imperative to know how the watershed behaves under different environmental conditions, parameter spatial variability, input aggregation and associated biases and how these differences are propagated to the solution. This knowledge enhances the forecast skills using distributed models such as: Wechsler [137]; Vieux et al. [127], Viglione et al. [134], Müller et al. [70], and Bloschl et al. [17].

Hydrologic parameters play an important role in the hydrologic prediction where high slope exist, and where soil as well as land use characteristics change over short distances. Hydrologic models average the hydrologic parameters and topographic characteristics in lumped, semidistributed and distributed models to simplify or reduce computational time. In addition, calibrations are usually limited to the watershed outlet, hence, not producing accurate flood prediction within the subwatershed's internal outlets.

Loss of accuracy occurs in flood prediction with topographic and parameters aggregation, however, how much loss of accuracy can we expect? Limited number of studies have evaluated the effects of grid size on basin response and the prediction of discharge in tropical environments and complex topography [18, 95, 118, 137, 139]. Therefore, the research in this chapter will investigate these aspects as they are related to model calibration and flood prediction.

The hydrologic model used in this research is V*flo*™ (for convenience in this dissertation V*flo*™ will be referred to as Vflo), a fully distributed hydrologic model [118, 124–126]. Vflo uses the finite element numerical method to resolve overland and channel flow. The Green Ampt equation is used to represent rainfall infiltration though the soil [80]. The digital revolution in geospatial data has helped to promote the development of physically based models capable of producing excellent results in flood prediction at internal basin points.

To understand the system predictability, authors conducted various experiments within a small subwatershed laboratory (test-bed) covering a 4×4 km^2 Geostationary Operational Environmental Satellite (GOES) pixel. This "real world" laboratory has a rain gauge network with a resolution well below that of the NWS radar products; a stage elevation station at the outlet; high topography resolution information (Digital Elevation Model raster map, DEM 10×10 m^2), remotely sensed data (e.g., LandSat Thematic M) and several field measurements to represent the channel geometry. The test-bed subwatershed is located in Western Puerto Rico and belongs to the Río Grande de Añasco watershed. To establish a flood alarm system in the region of the study area, it is necessary to know the performance and the prediction limits associated with the small subwatersheds.

1.2 JUSTIFICATION

A study which considers different input (rainfall) resolutions, parameter aggregation effects and hydrologic model resolutions, at scales lower than the current radar products, has not been conducted in Puerto Rico or anywhere. With the new emergent radar technologies, it is necessary to recommend to the hydrology community which grid size is necessary to capture the spatial variability of rainfall and hydrologic model that generate reliable flood prediction. The prediction limits related to this input grid size and, at the same time have a cell size that minimizes the computational time for real-time applications.

The grid size and the watershed response are interrelated. Therefore, it is imperative to know the combination of grid sizes needed to produce reliable results within the study area and to know the probabilistic distribution function (PDF) of flow peaks, time to peak and runoff volume associated with each resolution. The optimal grid size is defined as the largest grid size which will produce reliable results, beyond which flood prediction accuracy degrades.

The time required to run the model in real-time operation mode is critical. Therefore, the grid size should decrease the computational time, while maintaining sufficiently accurate results. An up-scaling evaluation of rainfall and hydrologic parameters consist in the creation of a high-resolution hydrologic model, and then increasing the grid size to produce incrementally coarser resolution maps of each parameter and input, resulting in different output responses. These hydrologic responses will be compared in terms of their probability distribution functions (PDFs) to observed values. A decision can be made in terms of which aggregation technique should be used to aggregate the data and which parameters will be used in the evaluation at small scales.

1.3   RESEARCH QUESTIONS

Real time hydrologic predictions require estimation of stream stage, peak flow, time to peak, and storm volume with high reliability. To obtain reliable estimates it is necessary to know and understand the predictability and prediction limitations of the system.

The general objective of this research was to evaluate the hydrologic predictability of flood predictions in complex terrain located at Mayagüez Bay drainage basins due to rainfall inputs and hydrologic model resolutions. To identified representative parameters at each scale that will enhance the flood prediction when the modeler uses different grid size resolution inputs within the distributed hydrologic models.

Three basic research questions (RQ) addressed in this research are summarized below and were based on a workshop on "Predictability and Limits to Prediction in Hydrologic Systems by the National Research Council [28]" and suggestions made by several investigators in this field [36, 37, 128–133].

RQ1.   How flow prediction is affected by the spatial variability of point rainfall at scales below that of the typical resolution of radar-based products?

The error propagation due a rainfall spatial resolution in the distributed models has been a goal in the hydrologic community in recent years. Different studies that have been conducted have been done at scales courser or same than resolution of the radar rainfall products using distributed models [26, 38, 39, 120] or using lumped model [12].

The accuracy of current precipitation estimates over a basin must be known; and moreover, the accuracy of these estimates must be improved before the uncertainty in hydrologic forecasts can be quantified and ultimately reduced. According to Droegemeier and Smith [27], hydrologic forecast uncertainty cannot be reasonably assessed until the uncertainty in the rainfall observations has been determined a priori. Entekahbi et al. [28] identified the uncertainty in model inputs as one of the major limitations to improved hydrologic predictability.

One important contribution will be to find the current rainfall product uncertainty over small watersheds. Also, evaluate how uncertainties due to quantitative precipitation estimates at different resolutions (below 2 km) from point rainfall are propagated though the hydrologic solution. By this means we can determine which rainfall resolution is required to encompass the rainfall variability and produce the least uncertainty and highest accuracy for flood predictions at scales below radar products and small subwatersheds.

The Collaborative Adaptive Sensing of the Atmosphere (CASA) project has instrumented a 4×4 km^2 area with a network of 28 rain gauges, producing high spatial rainfall resolution with the objective to test and validate CASA radars. Inside the pixel a small subwatershed was delineated and instrumented with a pressure transducer to measure stage at a determined cross section. The small area was named Test Bed Subwatershed (TBSW) and serves as a field laboratory to test how the uncertainty due to rainfall resolution input propagates though the distributed hydrological model to the streamflow prediction.

RQ2. *How does parameter and hydrological model resolution affect the model's predictive capabilities and the errors of the hydrologic system?*

To develop a real time hydrologic model, a coarse grid size resolution is desirable in order to minimize computational time. However, this choice could have an important impact on the hydrologic simulation, because the calibration is grid-cell size dependent. The effects observed in the grid size aggregation are flattening of the slope and shortening of the drainage length, changes in flow direction, channel and overland cells and smoothness of the soil parameters and roughness. Both effects can be compensated for or reduced depending on the topographic characteristics of the basin and the methods used to calculate them [20, 77, 102, 118].

Mountainous areas with large slopes are more sensitive to digital elevation model resolution. The resolution of the terrain model needed to capture the basin properties is the same for slope as it is for other parameters such as hydraulic roughness derived from land use obtained from satellite remote sensing and soil properties. Understanding the influence of resolution and parameter aggregation on the hydrologic model would enhance the model prediction. This will be accomplished using the highest resolution data available and then producing coarser resolution maps of each parameter though up-scaling (various methods could be tested here), and evaluate how the coarser resolution degrades the solution obtained at the finest resolution. Authors hypothesize that the finer hydrologic model resolution ensemble will have the best flow prediction behavior. However, this model is not operational for future flash flood forecasting. The goal is to find a practical grid size resolution for real time applications and address reliable results at small watersheds.

RQ3. *Would the assumptions developed for the small scale enhance the hydrologic predictability at larger scales?*

The hypothesis formulated is that if we can enhance the flood forecasting in small subwatersheds than we can enhance the flood forecasting at larger scales, where all major mountainous basins are composite of similar subwatersheds that have similar slope conditions, land use coverage

and soil properties. Lessons learned in this study about the small watershed's behavior could be applied to watersheds of major sizes where the
cost of using high-resolution data could result in better flood forecasting. However, if it is necessary to apply coarse resolution data to large
scale, real time applications, the predictability limits could be known a
priori. Recommendations related to which terrain and rainfall grid sizes
and parameter estimations to use in the distributed hydrologic model will
be available, and will be tested in watersheds of major size. Only a few
rain gauges and NEXRAD rainfall estimates are provided to major areas.

1.4   OBJECTIVES

The specific objectives of this study in Part I of this book, required for the
achievement of the major research goal and the research questions are:

a. Configure a hydrologic distributed model for the Mayagüez
    Bay Drainage Basin (MBDB) and extract a small subwatershed
    (TBSW) having similar slope characteristics to the MBDB subwa-
    tersheds, for the purpose of performing detailed studies.
b. Analyze the MBDB hydrologic model sensitivity in the flow
    response due to propagation of parameter and rainfall perturbations
    using spider plots and relative sensitivity analyzes.
c. Quantification of MBDB hydrologic model flow response due to
    two rainfall interpolation methods and radar sources.
d. Evaluate the rainfall detection accuracy of the current radar prod-
    uct (multisensor precipitation estimator, MPE) at scales below
    2 km using a high-density rainfall network.
e. Evaluate ensemble behavior for rainfall resolutions exposed to
    uncertainties in parameter quantifications and hydrologic model
    resolutions.
f. Evaluate ensemble behavior of hydrologic model resolutions due
    to propagation of parameter uncertainties and rainfall resolutions.

1.5   SUMMARY

The research study on "*Flood Prediction Limitations in Small Watersheds*"
is presented in detail in Chapters 1–9 of this book. The general objective of

this study was to evaluate the hydrologic predictability of flood predictions in complex terrain located at Mayagüez Bay drainage basins in Puerto Rico due to rainfall inputs and hydrologic model resolutions. To identify representative parameters at each scale that will enhance the flood prediction when the modeler uses different grid size resolution inputs within the distributed hydrologic models.

Three basic research questions (RQ) addressed in this research were based on a workshop on "Predictability and Limits to Prediction in Hydrologic Systems by the National Research Council [28]" and suggestions made by several investigators in this field [36, 37, 128–133]. These questions were: *RQ1. How flow prediction is affected by the spatial variability of point rainfall at scales below that of the typical resolution of radar-based products? RQ2. How does parameter and hydrological model resolution affect the model's predictive capabilities and the errors of the hydrologic system? RQ3. Would the assumptions developed for the small scale enhance the hydrologic predictability at larger scales?*

CHAPTER 2

FLOOD PREDICTION LIMITATIONS IN SMALL WATERSHEDS: A REVIEW[1, 2]

ALEJANDRA M. ROJAS-GONZÁLEZ

CONTENTS

2.1 QUANTITATIVE PRECIPITATION ESTIMATES

A major source of error in hydrologic models is the poor quantification of the areal distribution of rainfall, typically due to the low density of rain gauges. A rain gauge located at a single point may not represent an extensive area, with only one value. The spatial distribution of rainfall can have a major influence on the corresponding runoff hydrograph, errors may occur in the resulting hydrograph when the spatial pattern of the rainfall is not preserved. These errors will be magnified for intense, short duration

---

[1] This chapter is an edited version from *Alejandra María Rojas González, 2012. Flood prediction limitations in small watersheds with mountainous terrain and high rainfall variability. Unpublished PhD Thesis for Department of Civil Engineering and Surveying, University of Puerto Rico – Mayagüez Campus.*

[2] Numbers in brackets refer to the references at the end of this book.

and localized events especially in areas of high topographic variability subject to convective storms [143].

Rain gauges themselves may produce errors, a major source of error being from turbulence and increased winds around the gauge, affecting precipitation quantification in events where the wind is an important factor (e.g., hurricanes). Nevertheless, the rainfall measured in a gauge station is generally assumed to be the most reliable measurement of rainfall, but when measurements are extrapolated to the entire basin for hydrologic models, the rainfall has a great uncertainty and can affect the watershed response. Bevan and Hornberger [13] have stated, "... *an accurate portrayal of spatial variation in rainfall is a prerequisite for accurate simulation of stream flows*".

Investigators have used mean areal precipitation as calculated by Thiessen polygons [115, 143], and interpolation methods (Spline, Inverse Distance Weights, and Krigging and polynomial surface). But all of these methods are limited by the number of rain gauges.

Ball and Luk [7] studied the accuracy and reliability of hydroinformatic tools (e.g., GIS) for modeling the spatial and temporal distribution of rainfall over a catchment. They found that using spline surfaces with a geographic information system produced robust and accurate estimates of rainfall and enable real-time estimation of spatially distributed patterns.

Currently, sophisticated methods attempt to fill gaps between rain gauges, by sensing the atmosphere with remote sensors like the spaceborne Tropical Rainfall Measuring Mission (TRMM), the U.S. National Weather Service's (NWS) Next Generation Radar (NEXRAD), the National Oceanic and Atmospheric Administration's (NOAA) Hydro-Estimator (HE) algorithm [91], the satellite precipitation estimation/radar rainfall merging algorithm of the NOAA-CREST Group at City University of New York [62] and the MPE [57, 60, 92]. The HE uses data from the GOES geostationary satellite to estimate rainfall, and has, for example, an approximate pixel size of 4×4 km^2.

These quantitative precipitation estimation (QPE) techniques are evaluated and adjusted or calibrated using existing rain gauges, however, these adjustments depend on the rain gauge density and their spatial distribution [47]. Studies that have compared radar and rain gauge–derived rainfall documented large discrepancies among various investigators [6, 64, 144].

In order to address the need to obtain more rainfall estimates for basin analysis, in 1997 National Weather Service (NWS) put into operation the WSR-88D Next Generation Radar (NEXRAD) in the United States of America (USA). NEXRAD radar enhances coverage with a 1 degree × 1 km base resolution. Since 1999, NEXRAD has been used by the NWS to estimate rainfall in Puerto Rico. The NEXRAD facility is located near the City of Cayey at 860 m above mean sea level and at approximately 120 km from Mayagüez. The radar measures reflectivity in decibel (dBZ) and uses empirically derived Z-R relationships to transform reflectivity to rain rate. The Marshall and Palmer [63] equation is the default Z-R relationship employed by the WSR-88D and is described by the following empirical power law:

$$Z = aR^b \tag{1}$$

where Z is the reflectivity in decibels (dBZ) and R is the rain rate in mm/h; a and b are nonlinear regression coefficients and their respective values depend on the type of precipitation.

The coefficients depend on location, season, rain type, drop size distribution, and are event dependent. Battan [8] presents more than 50 Z-R relationships. Currently there are at least five different relationships depending on climatological zones approved by the NWS. For example for a convective rainfall, a and b values are 300 and 1.4, respectively. Similarly, under tropical conditions, values of 250 and 1.2, respectively, have been used and for a warm stratiform rainfall values of 200 and 1.6 are used.

The default Z-R relationship used in Puerto Rico is the convective type and is not representative of tropical rain events due to the drop size distribution (smaller rain drops than convective with fewer and larger rain drops). It is necessary to define a maximum precipitation rate threshold for decibels above 51, because Eq. (1) with the tropical coefficients can produce nonsensical rain rates. High dBZ are due to possible hail formations or very heavy precipitation or extreme winds, which also may be produced by thunder and lightning, and wet ground returns. The radar default setting is 4.09 inches/h and if rainfall rates are greater, a deep warm layer exists. Therefore, warm rain processes govern, which is typical of tropical events [61]. Operationally the Z-R relationship should be changed to the tropical equation and the maximum precipitation rate threshold changed to 6.00 inches/h.

Vieux and Bedient [121, 122, 129] found an improved Z-R relationship comparing slopes of the best-fit regression lines of each Z-R relationship to daily rain gauge accumulation. With the current Z-R relationship used in Puerto Rico, NOAA has reported low estimates of accumulated rainfall by the radar as compared to gauge accumulations.

The MPE algorithm is a product of NEXRAD, and recently has been used to improve quantitative precipitation estimates [58, 62]. MPE is based on the Digital Precipitation Array (DPA) product (hourly and 4×4 km^2 resolution) and performs a mean field bias correction over the entire radar coverage area, based on (near) real-time hourly rain gauge data [92, 93]. The MPE is mapped onto a polar stereographic projection called the Hydrologic Rainfall Analysis Project (HAP) grid. This data is often used in hydrologic modeling, availing the bias correction made by the MPE algorithm. Nevertheless, in long-term hydrologic simulations and watersheds with small numbers of rain gauges, a bias verification would be evaluated, because the bias quantification has a high variability over the radar coverage area and time [47, 78, 79] affecting the hydrologic calibration and validation.

Gourley and Vieux [38] developed a method for evaluating the accuracy of Quantitative Precipitation Estimates (QPE) for isolated events. A hydrologic approach to QPE evaluation may also become complicated because model parameters can be judiciously adjusted or calibrated to account for errors in model inputs. Systematic biases, which are originally present in the model inputs, can be mitigated or corrected in order to yield accurate streamflow forecasts.

Probabilistic calibration methods exist, such as the generalized likelihood uncertainty estimation (GLUE) used by Beven and Binley [14], to compute the probability that a given parameter set adequately simulates the observed system behavior. Furthermore, it was suggested by Freer et al. [33] that the GLUE technique should be expanded to include the uncertainties associated with different rainfall inputs. Extension of the GLUE provides a consistent methodology to independently evaluate the hydrologic response to each input.

Georgakakos [37] expressed the need of future research in the context of short-term hydrologic forecasting with QPF driven distributed hydrologic models which include:

a. Development of high-resolution reliable QPF, especially in mountainous areas.
b. Sensitivity analysis of distributed models with operational data to assess the relative importance of parameter uncertainty and QPF hydrologic models that include characterization of the errors in distributed QPFs.

2.2 PHYSICALLY-BASED DISTRIBUTED HYDROLOGIC MODELS

The term physics-based model means that conservation of mass in combination with momentum and/or energy is employed to compute hydrologic fluxes. Vieux and Moreda [125] indicated that the goal of distributed modeling of streamflow is to better represent the spatial-temporal characteristics of a watershed governing the transformation of rainfall into runoff that relies on conservation equations for the routing of runoff though a distributed representation of a watershed.

The term "physics-based or physically based distributed (PBD) models" includes such models as *Vflo* [122]; Vieux et al. [119, 122]; CASC2D [55, 56, 74]; Systeme Hydrologique European (SHE) [1, 2] and the Distributed Hydrology Soil Vegetation Model (DHSVM) [141]. PBD models are well suited to simulating specific events at locations where streamflow records may not exist.

Conceptual rainfall-runoff (CRR) models simulate runoff generation by a variety of conceptual parameters and route the runoff using unit hydrographs to an outlet. CRR models are inherently nonphysics based and lump parameters at the basin or subbasin level. CRR models include Precipitation-Runoff Modeling System (PRMS) by Leavesley et al. [59], the Sacramento Soil Moisture Accounting Model (SAC-SMA) [21], and the HEC–HMS model (Hydrologic Engineering Center) [53, 54]. CRR models differ from event-based models, simulating continuous cycles of rainfall and runoff. The CRR models breakdown the hydrologic cycle into a series of reservoirs that represent physical phenomena such as infiltration, runoff, etc. [125].

Physics-based models use conservation of mass, momentum, and energy equations to represent hydrologic processes, whereas conceptual models use empirical relationships together with buckets to represent component processes. Moore and Grayson [68] described an array of physics-based models that capitalize on digital models of elevation, GIS and remotely sensed (GIS/RS) geospatial data.

The model used in this research is a fully distributed, physics-based hydrologic model named *Vflo* [124, 127] that derives its parameters from soil properties, land use/cover, topography, and can obtain input from radar or multisensor precipitation estimates. *Vflo* incorporates routing of unsteady flow though channel and overland elements comprising a drainage network.

The following *Vflo* description and mathematical formulation was obtained (in some cases verbatim) from Vieux and Vieux [125], who stated that the model uses the kinematic wave analogy (KWA). The KWA has better applicability where the principal gradient is the land surface slope. Thus in almost all watersheds except for very flat areas, the KWA may be used. The simplified momentum equation and the continuity equation comprise the KWA. One-dimensional continuity for overland flow resulting from rainfall excess is expressed by:

$$\frac{\delta h}{\delta t} + \frac{\delta(uh)}{\partial x} = R - I \tag{2}$$

where, R is rainfall rate; I is infiltration rate; h is flow depth; u is overland flow velocity; t is the time and x is the distance.

In the KWA, the bed slope is equated with the friction gradient. In open channel hydraulics, this amounts to the uniform flow assumption. Using this fact together with an appropriate relation between velocity, u (m/s), and flow depth, h (m), such as the Manning equation, we obtain the velocity for very wide-open channel and metric system:

$$u = \frac{S_0^{\frac{1}{2}}}{n} h^{\frac{2}{3}} \tag{3}$$

where, S_o (m/m) is the bed slope or principal land surface slope, and n is the hydraulic roughness called Manning's coefficient.

Velocity and flow depth depend on the land surface slope and the friction induced by the hydraulic roughness. For channel flow, Eq. (2) is written with the cross-sectional area A instead of the flow depth h:

$$\frac{\delta}{\delta} \quad \frac{\delta}{\delta} \tag{4}$$

where, Q (m³/s) is the discharge or flow rate in the channel, and q is the rate of lateral inflow per unit length in the channel. Combining Eqs. (3) and (4), we get:

$$\frac{\delta h}{\delta t} + \frac{s^{\frac{1}{2}}}{\beta n} \frac{\delta h^{\frac{5}{3}}}{\delta x} = \gamma R - \alpha I \tag{5}$$

where, the three scalars α, γ and β are the multipliers for the values contained in the spatially variable parameter maps according to the Ordered Physics-based Parameter Adjustment (OPPA) calibration method. Differential application of the roughness scalars (βn) to channel and overland are used (βc for channel and βo for overland).

Overland flow is modeled with Eqs. (2) and (3), and channel flow with Eq. (4), and appropriate form of the Manning uniform flow relation in Eq. (4) using the finite element method.

Digital maps of soils, land use, topography and rainfall rates are used to compute and route rainfall excess though a network formulation based on the Finite Element Method (FEM) computational scheme described by Vieux [116] and Vieux et al. [117]. Special treatment is required to achieve a FEM solution to the KWA over a surface with spatially varying roughness, slope, or other parameters. Vieux et al. [117] presented such a solution using nodal values of parameters in a finite element solution. This method effectively treats changes in parameter values by interpolating nodal values across finite elements.

Vieux [122] and Vieux et al. [119] described the development of a rainfall-runoff model based on a drainage network comprised of finite elements. The advantage of this approach is that the kinematic wave analogy can be applied to a spatially variable surface without numerical difficulty introduced by the shocks caused by noncontinuous parameter variation that would otherwise propagate though the system. The finite element methodology results in execution times that are fast enough to allow real-time computation before the next radar update.

Accounting for unsteady flow in mild slopes, *Vflo* allows a looped rating curve for channel elements. Essentially, the acceleration (deceleration) induced by the rising (falling) limb of the hydrograph is accounted for though the Jones Formula [52]. In mild slope hydraulic conditions, looped rating curves may cause important effects when maximum flow rate is observed. *Vflo* incorporates both distributed runoff generation, and routing of unsteady flow though channel and overland elements [125].

Vieux and Bedient [121, 124] used spatial resolution of radar rainfall as input to a distributed model which affected prediction error. Also, Vieux and Imgarten [132] studied the scale-dependent propagation of hydrologic uncertainty using high-resolution X-band radar rainfall estimates for watershed areas less than about 20 km^2. Results of experiments using historical radar events and including the tropical storm Allison indicated that accurate rainfall-runoff predictions in real time are possible and useful for site-specific forecast in Houston, TX. They found that the achievable model accuracy with radar bias correction was approximately a mean absolute percentage error of 11.8% in peak discharge, 11.1% in runoff depth and average difference in arrival times of 12 min at the Main Street gauge with a drainage area of 260 km^2.

The complex interaction of input with drainage network presents challenges to the design of storm-water drainage infrastructure, the management of flooding, flood mitigation, and real-time forecasting of multiscale urban drainage systems with multiscale inputs [131].

2.3   CALIBRATION PROCESS

2.3.1   SENSITIVITY ANALYSIS

The classification of the sensitivity analysis methods refers to the way that the parameters are treated. Local techniques concentrate on estimating the local impact of a parameter on the model output. This approach means that the analysis focuses on the impact of changes in a certain parameter value (mean, default or optimum value). Opposed to this, global techniques analyze the whole parameter space at once. Global sampling methods scan in a random or systematic way the entire range of possible parameter values and possible parameter sets. The sampled parameter sets can give the user a good idea of the importance of each parameter. These in turn can be used to quantify the global parameter sensitivity or the uncertainty of parameters and outputs.

2.3.2 CALIBRATION OF DISTRIBUTED MODELS

Vieux and Moreda [126] developed an OPPA procedure for a distributed model. The OPPA calibration process involves estimating the spatially distributed parameters from physical properties, assign channel hydraulic properties based on measured cross-sections, study the sensitivity of each parameter, and find the optimum parameter set that minimizes the respective objective function. Runoff depth should be adjusted first, followed by timing and peak flow and readjust hydraulic conductivity if necessary to account for changes in infiltration opportunity time. The Vflo model does not simulate base flow directly, only direct runoff. It can be taken in account by assigning a fixed value to channel cells for one simulated event. For long-term analysis, it is necessary to quantify the base flow using known methodologies [43, 94] and subtract it from the observed hydrograph to compare with direct runoff simulated by the *Vflo* model.

The agreement between the observed and simulated runoff depth, time to peak and peak flow may be expressed in terms of a bias or spread. The bias indicates systematic over or under prediction. The departure, whether expressed as an average difference, percentage error, coefficient of determination, or as a root-mean-square error, serves as a measure of the prediction accuracy.

McMichael et al. [66] calibrated a distributed physically based hydrologic model (MIKE-SHE) in California and estimated uncertainty. They used the GLUE methodology for model calibration, testing and predictive uncertainty for estimating monthly streamflow. The catchment in Central California was 34 km^2 in area and the model grid size was fixed at

270×270 m². The Monte Carlo simulation was used to randomly generate one thousand parameters sets for a 20-year calibration period encompassing variable climatic and wildfire conditions. Many studies have demonstrated the difficulties that arise in identifying, calibrating and validating physically based hydrologic models. Such difficulties stem from uncertainties in model structure, boundary conditions, and catchment parameterization, as well as errors in inputs and observed variables.

The GLUE methodology [14, 15] explicitly recognizes the coexistence of alternative parameter set and models and it provides a suitable framework for model calibration and uncertainty estimation under nonuniqueness. The nonuniqueness recognizes the existence of several set of parameters and structures that would produce good agreement with the observed data, and satisfy the calibration. With the limited measurements available and the application of a distributed hydrological model it may not be possible to identify an optimal model. Implementing GLUE requires making Monte Carlo simulations using a large number of parameter sets, assessing the relative performance of each set by comparing model estimates with observed data, and retaining only those parameter sets that provide behavioral (acceptable) predictions. The relative performance of each parameter set is evaluated on the basis of a likelihood measure calculated by comparing model predictions with observed data. A parameter set is classified as behavioral if the corresponding likelihood value is equal to or greater than a specified threshold value. Parameters sets that do not meet this criterion are rejected as nonbehavioral.

The final step in the GLUE procedure is to establish predictive uncertainty bounds for comparison with observed values. First, the set of behavioral likelihood values is rescaled to archive a cumulative sum of unity by dividing each value by the sum of the likelihood values. Next, behavioral model predictions for each time step are ranked in ascending order and each prediction is assigned to a user-specified bin. The rescaled likelihood values associated with the ranked predictions in each bin are summed to calculate the height of the corresponding bar in the density plot. A cumulative density plot is constructed by graphing the cumulative sum of the likelihood values versus the ranked model predictions. Typically, the 5th and 95th percentiles calculated at each time step are used to calculate the predictive uncertainty bounds over the period of observations. The GLUE based prediction limits the capture of uncertainly in model output associated with uncertainly in model parameterization.

GLUE provides a useful modeling approach for advancing beyond globally optimized, unique, parameter sets. Working within a framework of Monte Carlo-generated parameters sets allows modelers to explicitly recognize and quantify the effects of uncertainties on model prediction [66].

Sahho et al. [84] performed a calibration and validation of MIKE SHE in a flashy mountainous Hawaii stream. The model was calibrated with a single hydraulic conductivity value and produced consistent results with correlation coefficients greater than 0.7. In the sensitive analysis the Manning's roughness coefficient and the hydraulic conductivities (vertical and horizontal) of the saturated zone had the most pronounced effects in determining the shape of the flood's peaks.

Griensven et al. [41] made a global sensitivity analysis tool for the parameters of multivariable catchment models. An analysis of Monte Carlo simulations was conducted with statistical methods such as Kolmogorov–Smirnov (K-S) test [100] or with the computation of regression and correlation based sensitivity measures to define whether a parameter is sensitive [98]. An advantage of the method is the logical combination of calibration, identifiable analysis, and sensitivity and uncertainty analysis within a single modeling framework [113]. The method can be applied to problems with absolutely no probabilistic content as well as to those with inherent probabilistic structure. It has been widely used in catchment modeling, for assessing parameter uncertainty and input uncertainty, e.g., for rainfall variability.

The Monte Carlo method provides approximate solutions to a variety of mathematical problems by performing statistical sampling experiments on a computer [31]. This method performs sampling from a possible range of the input parameter values followed by model evaluations for the sampled values. An essential component of every Monte Carlo experiment is the generation of random samples. Techniques, such as the Latin – hypercube methodology, are also available for minimizing the number of required runs to reproduce the selected probability distributions of the input datasets [46]. These generating methods produce samples drawn from a specified distribution (typically a uniform distribution). The random numbers from this distribution are then used to transform model parameters according to some predetermined transformation equation.

2.4   FLOOD PREDICTION

In an attempt to determine flood occurrence, Birikundavyi et al. [16] used two approaches commonly used for the probabilistic analysis of extreme flood magnitudes that are based on the annual maximum series (AMS) and the partial duration series (PDS). In the AMS approach the highest flood peak in the year is used, while in the PDS approach all those events that exceed a specified value are used. In the study, the Poisson distribution and generalized Pareto distribution (GPD) were used to describe the occurrence of flood and the flood magnitudes. Two neighboring flood peaks were independent if (1) they are separated by at least seven days and (2) the flow between them drops below 50% of the smaller peak.

In the Brays Bayou watershed (334 km²) in south-west Houston Texas, Bedient et al. [10] developed a flood warning system using radar-based rainfall (NEXRAD) and delivery systems on the internet. During 1950–1960 the Army Corps of Engineers constructed a concrete and rip-rap lined channel to contain a greater than 100-years storm event with bankfull capacity, currently the same channel only can contain the 10 year design level due to increased urbanization. In this system HEC-1 is used to predict the flow at different interest points with known rainfall distribution and the results are modeled in HEC-2 to determine the maximum height of water in the channel. These two models are often used together for flood prediction and are the basis for calculating the Flood Alert System monograph used to translate rainfall rates into peak flow and levels. After, generating the system monograph, calibration was conducted with hypothetical storms.

The HCOEMALERT (Harris County Office of Emergency Management Automated Local Evaluation in Real-Time) exists within the Brays Bayou watershed with a high density of rainfall and flow gauges available real time via the internet [10, 11, 50]. Data received from these gauges can be used to predict possible flooding conditions and were used to calibrate the watershed HEC-1 model.

NEXRAD used with GIS can calculate the rainfall rates within the subwatersheds and to estimate rainfall rates from approaching storms and visualize the development of the storm. These are powerful tools for storm prediction and flood alert. Bedient et al. [10, 129] reported an excellent accuracy using HEC-1 and NEXRAD in several storms. However, the NEXRAD data was only used to track the storm.

Currently, *the next generation flood alert system* (FAS2) started its operation in 2004 with more than 30 storm events [30]. FAS2 uses available radar (NEXRAD) data coupled with real-time hydrologic modeling, and provides visual and quantitative identification of severe storms producing heavy rainfall, as well as a linkage between the rainfall and likelihood of flooding. The accuracy of the current FAS2 is adequate for regional events over a large basin (129 mi^2), but is lacking for events where the regional/local scale interactions, local scale precipitation, infiltration losses, or local hydraulics are important.

In the CASA Annual Report year 3, Volume II [22], three projects were cited that are in development which are employing state-of-the-art techniques. In the S22 project, it uses rainfall data derived from radar images to run real-time, physically based distributed models for flood prediction and generation of flooding maps. This project explores the drainage density in an urban area, because it has been demonstrated in FAS that a small urban watershed could not predict flow with sufficient accuracy with the current Vflo model, when the area was classified as overland flow.

Project S23 is concerned with testing different QPE resolutions derived from radar and the impact in flow at different basin scales with the same grid size resolution. Project S24 is developing a *Vflo* model that incorporates a secondary drainage system and evaluating the methodology in Harris Gully (FAS's urbanized watershed). A distributed pipe network linked to topography is a unique combination of new urban hydrologic models. All these projects are guided to enhance the accuracy in flood prediction especially at small watershed scales.

Making predictions in real-time with a hydraulic model is difficult because of inaccuracies in model parameters, rainfall input inaccuracy, or unknown upstream flow rates. Real-time systems for mapping expected areas of inundation require input of flow rates from other sources to generate inundated areas using sophisticated 2-D hydrodynamic models [140]. Even the inflow between river gauging stations requires some model estimation of watershed response in the intervening areas. Upstream gauging points and rainfall-runoff models are viable sources of real-time flow information. Both lumped and physics-based distributed rainfall-runoff models may be used for this purpose [11].

Georgakakos [36] studied the theoretical basis of developing operational flash flood guidance systems using analytical methods. The Sacramento soil moisture accounting model is used operationally in the United States to produce flash flood guidance estimates of a given duration from threshold runoff estimates. The study attempted to: (a) shed light on the properties of this model's short-term surface runoff predictions under substantial rainfall forcing; (b) facilitate flash flood computations in real time.

Various characteristics of the flash flood guidance to threshold runoff relationship are discussed and considerations for real-time application are offered. Uncertainty analysis of the threshold runoff to flash flood guidance transformation is also performed.

Vieux et al. [127] in collaboration with Taiwan government agencies and the United States Government began a program initiative for the research and development of a flood alert and water resources management system to unify monitoring and prediction of floods within a single system in Taiwan. Enhancing the accuracy and efficiency of information disseminated from the central government to the public, and to regional and local water management and emergency response agencies is the major goal of this project. A limited sensitivity analysis was conducted. Knowing which parameters generate a greater response in stage or discharge; helps to identify where efforts should be expended to improve parameter specification.

Vieux et al. [123, 124] developed a proposal for Arizona State to use a sophisticated hydrologic modeling approach coupled with QPe-SUMS. This model can help to: (1) manage reservoir operations, (2) minimize losses though spills, and (3) predict flood levels in selected basins. The authors emphasize the need to perform a flood hazard analysis a priori to the modeling.

The U.S. Army Corps of Engineers [104, 105] define the *maximum potential warning time*, as the response time after initiation of the flood-producing rainfall and is related to the arrival time of the peak stage or discharge, and is the interval during which mitigating responses can reduce property damage, loss of life, or business interruption.

CHAPTER 3

FLOOD PREDICTION LIMITATIONS IN SMALL WATERSHEDS: HYDROLOGIC MODEL CONFIGURATION AND SLOPE ANALYSIS[1,2]

ALEJANDRA M. ROJAS-GONZÁLEZ

CONTENTS

---

[1] This chapter is an edited version from, "*Alejandra María Rojas González, 2012. Flood prediction limitations in small watersheds with mountainous terrain and high rainfall variability. Unpublished PhD Thesis for Department of Civil Engineering and Surveying, University of Puerto Rico – Mayagüez Campus*".

[2] Numbers in brackets refer to the references at the end of this book.

3.1   INTRODUCTION

In this research study, the configuration for the MBDB model was developed using available data for soils, land use, digital elevation models and field measurements. This model will be used for uncertainty analysis, rainfall tests and posterior flood alarm predictions (not addressed in this research). Therefore, the TBSW model set up was conducted by extracting data from the MBDB model. A slope analysis was developed according to an aggregation method to be used in the up-scaling experiment, without loss of slope information for mountainous subwatersheds. Additionally, an evaluation between different evapotranspiration methods was developed to quantify the uncertainty associated with this term.

The hydrologic model used in this study is *Vflo* [127], which is capable of ingesting distributed radar rainfall data. *Vflo* is a finite element model and the equations are used to solve overland and channel flow.

The configuration of the proposed physically based distributed model used in this study was based on products described for the Mayagüez Bay Watershed and TBSW as well, such as soils, land use and digital elevation model maps. Generally, to create both high-resolution models, it is necessary to derive the topographic characteristics from a digital elevation model with high-resolution. For this purposes we used the digital elevation model quadrangles derived from the base map data of the "Center for Municipal Tax Revenues of Puerto Rico" by its acronym in Spanish [25: xyz mass points, ridgelines, road cuts, and hydrographic features]. The CRIM data were collected by AEROMETRIC, Inc. Ground control eastings, northings and elevations were surveyed by RLDA Surveying and Mapping of San Juan, Puerto Rico. The elevation maps were developed by photo-triangulation with a root mean square error of ground-control residuals of 0.6 m for vertical control elevation coordinates and root mean square error of airborne-GPS exposure-station residuals of 0.184 m for vertical control elevation coordinates.

Most of the input data for the *Vflo* model was prepared using ArcGIS 9.3 and Arc Hydro Tools. The basin and river characteristics were extracted from the 7.5-minutes series topographic maps from USGS, 30 m × 30 m² digital elevation model (DEM) quadrangles and from the digital elevation model at 10 m spatial resolution from CRIM.

The Green Ampt infiltration model is used by the distributed hydro-logical model to calculate the initial abstractions due to infiltration and runoff produced by rainfall. The parameters are derived from soil characteristics assigned to the SSURGO soil classification maps, digitally available (Figure 3.3). Values of soil suction at wetting front (ψ), saturated hydraulic conductivity (K_s), effective porosity, soil depth and initial degree of soil saturation (θ) were obtained from the literature [94, 132, 133], field measurements [USDA, 106–109] and computations using the percent of sand and clay, soil bulk density and percent of organic matter in combination with the *Soil Water Characteristics Hydraulic Properties Calculator* [85].

Vflo also requires soil depth (cm), initial abstraction (cm) and percentage of impervious area. Required channel data include base flow, roughness (Manning's n), channel and side slopes, and the infiltration parameters mentioned above. Overland flow properties include flow direction, overland slope and infiltration parameters.

3.2 STUDY AREA

3.2.1 MAYAGÜEZ BAY DRAINAGE BASIN STUDY AREA

The study area is located in the region of western Puerto Rico and has 819.1 km². The area includes three principal courses: Río Grande de Añasco, Río Guanajibo and Río Yagüez. Numerous hydrologic and hydraulic studies by the US Geological Survey (USGS) and the University of Puerto Rico have been conducted in this area [75, 82, 94, 135].

The area encompasses the municipalities of Mayagüez, Añasco, Las Marías, San Sebastián, Lares, Maricao, Yauco, Adjuntas, Sabana Grande, Cabo Rojo, San Germán and Hormigueros. Of these municipalities, Mayagüez has the highest population (89,080 habitants), followed by Cabo Rojo (50,917 habitants). The lowest population density is for Maricao with 6,276 habitants, according to the U.S. Census Bureau [103]. Changes in elevation vary from zero meters mean sea level in the coastal areas to 960 m in the mountainous areas, producing abrupt slope changes in short distances (Figure 3.1).

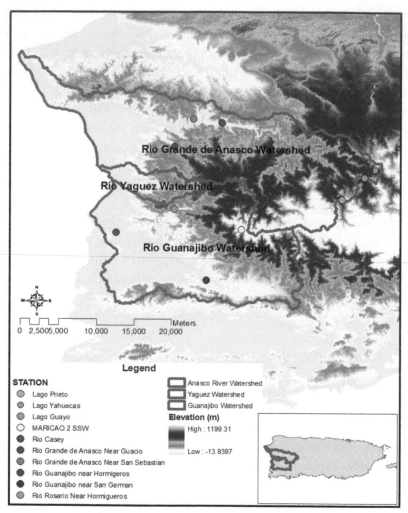

FIGURE 3.1 Digital Elevation Model (DEM); Río Guanajibo, Yagüez, and Grande de Añasco watersheds, rain gauges and flow gauging stations.

3.2.2 THE RÍO GRANDE DE AÑASCO BASIN

The Río Grande de Añasco basin (Figure 3.1) has an area of 370.36 km², including the reservoir lakes, tributary areas and river, which has a length of 64 km. Lakes Yahuecas, Prieto, Guayo and Toro were constructed by the Puerto Rico Water Resources Authority (PRWRA), presently the Puerto Rico Electric Power Authority, during the decade of the 50's. These were constructed to supply water to the Luchetti Lake for energy production and irrigation. According to Figueroa et al. [35], the area above Lago Guayo, Lago Yahuecas, and Lago Prieto dams contributes flow to the Río Grande de Añasco only during high floods. For the purpose of the present study it was assumed that the contribution of water from the Lago Guayo, Lago Yahuecas, and Lago Prieto sub watersheds to the Añasco watershed downstream of the lakes is not significant for regional water budget estimation [75]. Therefore, those sub watersheds were not included as part of the Añasco watershed in this study. The total lake drainage area is about 116.55 km^2 and was used as a boundary condition in the model.

The coastal plain associated with Río Grande de Añasco basin is characterized by an alluvial fan having an area of 41.5 km^2 and 0.08% average slope. The alluvial fan has a length of 15.6 km reaching a width of 8.8 kilometers at the coast shore [82].

According to FEMA [32], the estimated 100 years return period flood flows was 5,130 m^3/s (cms) and 3,797 cms for 50 years return period at the river's mouth. At USGS gauge No. 50144000 Río Grande de Añasco near San Sebastian, these were reported to be 4,078 cms for 100 years and 3,278 cms for 50 years return period. The major flood measured in that station was for Hurricane Georges in September 22, 1998, reporting a stage of 10.52 m (34.5 ft.) and peak flow of 4,587 cms, followed by Hurricane Eloise in September 16, 1975 with a stage of 10.33 m (33.9 ft.) and peak flow of 3,964 cms.

The station has different flood categories; the flood stage is 3.35 m (11 feet): a stage greater than 4.27 m (14 ft.) is a moderate flood and stages greater than 5.59 m (19 ft.) are categorized as major floods. The station shows that the river had been flooded in 30 one times since 1963 according to the records [73].

The Federal Emergency Management Agency (FEMA) performed a *Flood Insurance Study* (FIS) for the Commonwealth of Puerto Rico [32] in which regulatory peak flow values for the study basins were established. The Río Grande de Añasco FIS presents the magnitude and frequency of floods in accordance with the application of the U.S. Geological Service (USGS) regression equations for estimating peak flow on stream in Puerto Rico [111]. This report presented regression equations developed from gages sites having 10 to 43 years of records that can be used to estimate peak flows at ungagged sites or gaged sites with short periods of records.

The equations used the mean annual rainfall (MAR), the contributing drainage area (CDA) and the depth to rock (DR), as variables that govern the peak streamflow. The MAR was obtained from the Puerto Rico 1971–2000 Mean Annual Precipitation map developed by NOΛΑ [72], with the variations of rainfall across Puerto Rico calculated.

3.2.3   THE RÍO GUANAJIBO BASIN

The Río Guanajibo basin (see Figure 3.1) has an area of 328.9 km² and 38 km river length. The topography of the area is diverse, including mountains, foothills, and valleys. The predominant rocks in this area are serpentine and volcanic-related. The main tributaries are Río Rosario, Río Dagüey, Río Cain, Río Cupeyes, Río Cruces, Río Loco, and Río Viejo, and to the south exists relatively small tributaries. Major floods have been monitored in this basin since 1974, with the largest flood registered occurring in September 16, 1975 (Hurricane Eloise) with a reported peak flow of 3,625 cms and 8.7 m (28.54 ft.) stage elevation at the USGS 50138000 Río Guanajibo near Hormigueros station. In this location FEMA calculated a flow of 5,343 cms and 5,745 cms at the river's mouth for the 100 year return period. The 50 years return period flows were 3,637 at USGS station (50138000) and 3,896 cms at mouth [32].

The station has different flood categories; flood stage greater than 7.93 m (26 ft.) is categorized as a major flood, 6.7 m (22 ft.) is a moderate flood stage, 6.1 m (20 ft.) is the flood stage and at 4.88 m (16 ft.) is the stage at which action is required. The area had been flooded 20-four times since 1974 according to the records [73]. The percent annual chance recurrence intervals were developed using rainfall-frequency relationships presented in Technical Paper 42 (U.S. Department of Commerce, 1961) and an unit hydrograph was carried out using the HEC-1 computer program [USACE, 104].

The Río Rosario is a tributary of the Río Guanajibo and the subwatershed in this study is defined by the outlet point defined at the USGS 50136400 Río Rosario near Hormigueros station.

3.2.4   THE RÍO YAGÜEZ BASIN

The Río Yagüez basin (see Figure 3.1) has an area of 35.48 km², a river length of 20 km with average slope from 0.004% to 0.025% for the channelized river section at city of Mayagüez. Río Yagüez originates in the western slopes of the Cordillera Central and flows westerly into the Mayagüez Bay. The drainage basin is narrow, having a length-width ratio of approximately 10 to 1. In 1968, a flood protection project for the City of Mayagüez was initiated and the lower reach of the river was channelized to protect the city from floods. The channel has a capacity of 326 cms, but the maximum capacity of the channel at the PR Highway 2 Bridge is approximately 425 cms. To determine the discharges for the different percent annual chance floods in the basin reported in the FIS [32], a regional flood-frequency analysis [112] was used based on log-Pearson Type III analyzes of individual station records and regionalization using multiple regression techniques. The 100, 50 and 10-year return period flows at the mouth were estimated to be 770 cms, 595 cms and 292 cms, respectively [32].

Currently there are only four flow gauge stations with precipitation data and 2 river stage measurements (see Table 3.1 for the source and data type details). Nine flow gauge stations operated by the United States Geological Survey (USGS) exist within the study area (Figure 3.1):

- Three NOAA rain gauge stations;
- Two Soil Climate Analysis Network (SCAN) sites from the United States Department of Agriculture (USDA) Natural Resources Conservation Service (NRCS); and
- Four owner stations published at the underground web page (http://www.wunderground.com/US/PR/) [138].

The climate in the area is tropical, with moderate temperatures year round, and the mean high annual temperatures are 26.4 C in the mountains (Maricao 2SSW station) and 31.4 C in Mayagüez City station (Table 3.2). Table 3.2 presents a summary of the mean monthly average air temperatures and rainfall for five locations within the study area. Puerto Rico has a bimodal rainfall distribution in the wet season from April to November, with drier conditions in June and July; and a dry season from December to March.

The mean annual precipitation varies greatly across the study area due to the abrupt changes in elevation by the mountains causing wide variation in local wind speed and direction, which results in a sea breeze effect in the western area. Table 3.2 presents annual rainfall accumulations from 2463.8 mm for Maricao Fish and 1743.96 mm for Mayagüez City stations.

TABLE 3.1   Climatic and River Flow Stations Located Within the Study Area

Source	ID Station	Station Name	Lat.	Long.	Elev. (m)	Data
NOAA		Maricao 2 SSW	18.15	−66.98	863.4	Meteorclogical
NOAA		Hacienda Constanza	18.11	−67.05	146.3	Rain
NOAA		Mayagüez Airport	18.25	−67.13	11.6	Meteorological
NOAA		Maricao Fish Hatchery	18.16	−66.98	457.3	Rain
NOAA		Mayagüez City	18.18	−67.13	22.6	Rain, Temp
USGS	50131990	Río Guanajibo at Hwy 119 at San Germán	18.09	−67.03	45.0	Rain, Stage
USGS	50136400	Río Rosario near Hormigueros	18.17	−67.07	50.0	Rain, Stage, Flow
USGS	50138000	Río Guanajibo near Hormigeros	18.14	−67.15	2.2	Rain, Stage, Flow
USGS	50141500	Lago Guayo at Damsite near Castaner	18.21	−66.83	426.8	Rain, Stage
USGS	50142500	Lago Prieto near Adjuntas	18.19	−66.86	600.2	Rain, Stage
USGS	50146073	Lago Dagüey above Añasco	18.301	−67.13	40.0	Rain, Stage
USGS	50141100	Lago Yahuecas near Adjuntas	18.22	−66.82	426.8	Rain, Stage
USGS	50143930	Río Grande de Añasco at Bo. Guacio	18.28	−67.02	64.9	Rain, Stage
USGS	50144000	Río Grande de Añasco Near San Sebastián	18.285	−67.05	31.6	Rain, Stage, Flow
USGS	50145395	Río Casey above Hacienda Casey	18.25	−67.08	75.0	Rain, Stage, Flow

TABLE 3.1 Continued

Source	ID Station	Station Name	Lat.	Long.	Elev. (m)	Data
NRCS		Mayagüez TARS	18.217	−67.13	13.7	Meteorological
NRCS		Maricao Forest	18.15	−67.00	747.0	Meteorological
Wunder ground	MMGZP4		18.218	−67.16	0	Meteorological
Wunder ground	KPRMAYAG8	Miradero Mayagüez	18.23	−67.14	23.2	Meteorological
Wunder ground	KPRMAYAG7	Mayagüez	18.211	−67.14	0	Meteorological
Wunder ground	KPRMAYAG3		18.168	−67.15	48.8	Meteorological
Wunder ground	KPRSANGE3	Vivoni	18.083	−67.04	47	Meteorological

TABLE 3.2   Temperature and Precipitation for NOAA Stations Within the Study Area [72]

Station Name		Jan	Feb	Mar	Apr	May	Jun	Jul	Aug	Sep	Oct	Nov	Dec	Annual
Mayagüez City	High Temp (°C)	30.1	30.2	30.7	31.0	31.7	32.6	32.6	32.7	32.5	32.2	31.4	30.3	31.4
	Low Temp (°C)	17.9	17.7	18.1	19.2	20.3	20.9	21.1	21.0	21.2	21.1	20.2	19.1	19.8
	Rain (mm)	40.4	64.0	77.5	102.6	184.4	160.5	220.5	232.7	269.2	226.8	119.4	45.7	1744.0
	Days with Rain	6.7	6.7	8.3	10.9	14.2	13.6	15.6	17.2	17.1	16.1	11.8	7.9	146.1
Hacienda Constanza	Rain (mm)	48.5	67.1	80.3	120.7	200.2	195.6	247.1	253.7	279.4	242.6	138.4	34.0	1908.3
	Days with Rain	2.4	2.4	2.9	5.3	8.4	7.1	8.7	9.3	9.6	9.2	4.9	1.7	71.9
Maricao 2SSW	High Temp (°C)	24.8	25.2	25.7	26.2	26.7	27.8	27.6	27.8	27.4	26.8	26.2	24.8	26.4
	Low Temp (°C)	16.4	16.2	16.2	16.8	17.9	18.8	18.9	19.2	19.0	18.9	18.2	16.9	17.8
	Rain (mm)	76.5	95.3	134.1	172.5	239.5	159.5	216.4	287.0	348.0	378.5	236.0	86.6	2428.2
	Days with Rain	10.2	9.4	10.1	11.9	13.5	10.6	13.4	14.8	15.5	17.1	13.7	10.8	151
Mayagüez Airport	Rain (mm)	41.4	51.1	71.4	98.8	191.3	178.1	237.5	251.0	266.7	223.5	123.2	37.8	1771.4
	Days with Rain	6.9	6.4	8.4	11.5	16.6	15.5	17.4	19.3	18.7	18	12.2	8.4	159.3
Maricao Fish Hatchery	Rain (mm)	67.8	80.8	124.5	178.6	243.8	204.0	228.1	294.6	391.2	365.8	215.1	70.4	2463.8
	Days with Rain	5.5	5.3	6.9	10.2	11.3	8.5	11	13.3	15.4	15.7	10.5	7	120.6

3.2.5 SOILS CLASSIFICATION

A soil map describing the textural or soil class distribution is necessary to assign the values of the Green-Ampt infiltration parameters. The soil map was obtained from the Soil Survey Geographic (SSURGO) database for the Arecibo, Mayagüez, Lajas Valley and Ponce areas [USDA, 106–110] provided by the NRCS. Figures 3.2 and 3.3 depict the soil and textural classes occurring within the study area. The soil textures in the study area are: clay with 558.68 km^2 area, loam with 176.84 km^2, clay loam with 53.88 km^2, sand with 14.28 km^2, rock with 10.32 km^2 and gravel with 4.72 km^2. The SSURGO database provides additional information for each soil type, for example, bulk density, percent of sand and clay and soil depth. The soils series with a major presence in the area are Consumo (184.4 km^2), Humatas (132.9 km^2) and Mucara (78.9 km^2). The three soil types are classified clays for texture class, but have different infiltration

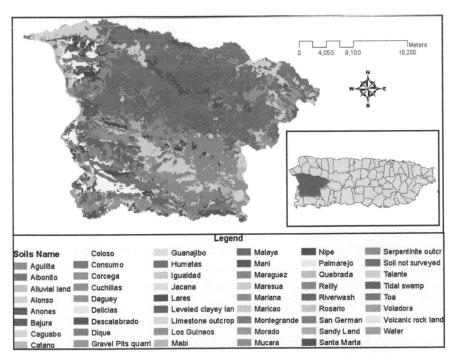

FIGURE 3.2 Soil map distribution for the study area. *Source:* SSURGO database, [USDA, 106–109].

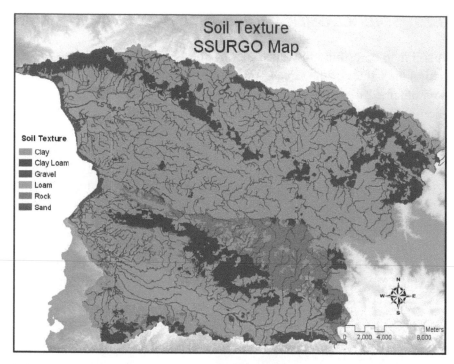

FIGURE 3.3   Soil texture for the study area, SSURGO map [USDA, 106–109].

capacities. Therefore, they are classified in the Hydrologic Soil Group as B for Consumo, C for Humatas and D for Mucara.

3.2.6   LAND USE CLASSIFICATION

To conceptualize the hydrologic model, it is necessary to obtain land use or land cover classes to assign roughness values and crop coefficients according to the classes. A digital map of the forest type and land cover was developed for Puerto Rico using LandSat enhanced Thematic images at 30 m resolution [51], applying a supervised classification approach. In total, 20-five classes were obtained from supervised classification (Figure 3.4). Prieto [75] reclassified the detailed classification into six major categories, grouping similar categories such as different forest types, shrub land, woodland or shade coffee.

The final land use classification is shown in Figure 3.5 and exhibits the predominant land use classification of forest, shub, wood land and shade

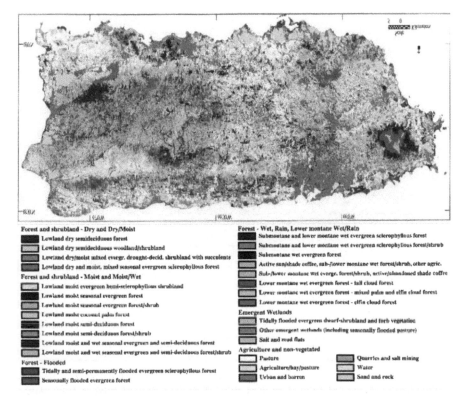

Forest and shrubland - Dry and Dry/Moist
- Lowland dry semideciduous forest
- Lowland dry semideciduous woodland/shrubland
- Lowland dry/moist mixed evergr. drought-decid. shrubland with succulents
- Lowland dry and moist, mixed seasonal evergreen sclerophyllous forest

Forest and shrubland - Moist and Moist/Wet
- Lowland moist evergreen hemi-sclerophyllous shrubland
- Lowland moist seasonal evergreen forest
- Lowland moist seasonal evergreen forest/shrub
- Lowland moist coconut palm forest
- Lowland moist semi-deciduous forest
- Lowland moist semi-deciduous forest/shrub
- Lowland moist and wet seasonal evergreen and semi-deciduous forest
- Lowland moist and wet seasonal evergreen and semi-deciduous forest/shrub

Forest - Flooded
- Tidally and semi-permanently flooded evergreen sclerophyllous forest
- Seasonally flooded evergreen forest

Forest - Wet, Rain, Lower montane Wet/Rain
- Submontane and lower montane wet evergreen sclerophyllous forest
- Submontane and lower montane wet evergreen sclerophyllous forest/shrub
- Submontane wet evergreen forest
- Active sun/shade coffee, sub-/lower montane wet forest/shrub, other agric.
- Sub-/lower montane wet evergr. forest/shrub, active/abandoned shade coffee
- Lower montane wet evergreen forest - tall cloud forest
- Lower montane wet evergreen forest - mixed palm and elfin cloud forest
- Lower montane wet evergreen forest - elfin cloud forest

Emergent Wetlands
- Tidally flooded evergreen dwarf-shrubland and forb vegetation
- Other emergent wetlands (including seasonally flooded pasture)
- Salt and mud flats

Agriculture and non-vegetated
- Pasture
- Agriculture/hay/pasture
- Urban and barren
- Quarries and salt mining
- Water
- Sand and rock

FIGURE 3.4 Map of Puerto Rico natural vegetation and land cover. (Reprinted from Helmer, E. H., Ramos, O., López, T. M., Quiñónez, M., & Díaz, W. (2002). Mapping the forest type and land cover of Puerto Rico, a component of the Caribbean biodiversity hotspot. *Caribbean Journal of Science, 38*(3/4), 165–183. With permission from the University of Puerto Rico at Mayagüez.)

coffee with an area of 529.16 km², followed by pastures with an area of 172.84 km² and Urban and barren land with 60.02 km². Preliminary, hydrologic model for the Mayagüez Bay basin area was configured using the Land use classification in Figure 3.5 provided by Prieto [75] and some analysis were developed using this data.

The second source of land use classification was provided by Puerto Rico Water Resources and Environmental Research Institute [PRWRERI, 76], who developed the project titled *Land Use Classification of the Mayagüez Bay Watershed, (Río Grande de Añasco, Río Yagüez, and Río*

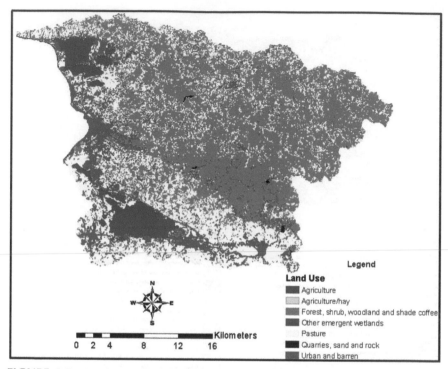

FIGURE 3.5   Land use classification at 30 m resolution from LandSat ETM, 2000. *Source:* Prieto, M. G., (2006). Development of a Regional Integrated Hydrologic Model for a Tropical Watershed.Master of Science Thesis, University of Puerto Rico at Mayagüez, PR.

Guanajibo Watersheds), supported by the Puerto Rico Environmental Quality Board (Figure 3.6). The sensor used for this classification was LANDSAT-7-TM satellite image from 2004 with 30 m resolution for a general land use classification with field visits verification as needed. Thirty-five classes were found in this product, where the most important area is covered by Forest low density (274.68 km²), fallow by Shrub and brush rangeland (253.05 km²), Forest high density (183.20 km²) and Urban or built-up land (103.71 km²).

3.2.7   TEST BED SUB WATERSHED

The *test-bed subwatershed* (TBSW) study area is located within the Río Grande de Añasco Basin, more specifically in the Río Cañas subwatershed

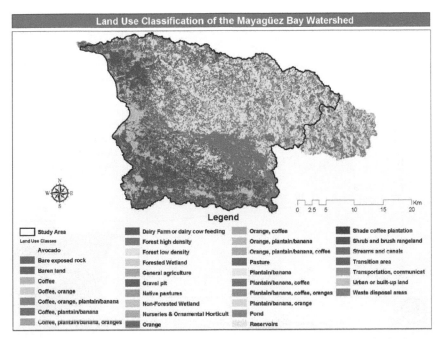

FIGURE 3.6 Land use classification of the Mayagüez Bay watershed, *Source:* PRWRERI, (2004). Land Use Classification of the Mayagüez Bay Watershed, Río Grande de Añasco, Río Yagüez, and Río Guanajibo Watersheds. Puerto Rico Water Resources and Environmental Research Institute (PRWRERI). Developed for the Puerto Rico Environmental Quality Board.

(Figure 3.6). In this study, the TBSW with an area of 3.55 km^2 is characterized and used for analysis purposes as a "field laboratory" to test the scale influence in the hydrologic prediction. The terrain elevation within the TBSW varies from 25.4 m (above mean sea level, amsl) to 305.7 m amsl, [25] (Figure 3.7). The area is characterized with large terrain elevation changes over small distances, with slopes varying from 0.265% to 91.96% (39.03% average slope). Therefore, the study area is classified as a mountainous sub watershed which is very typical of the Puerto Rican upland sub watersheds. Prior to this investigation, no rain or flow gauges were present within the area. Figure 3.7 shows the TBSW location within the Mayagüez Bay model, the color contoured terrain map and the rain gauge network installed and used in the study area for this research.

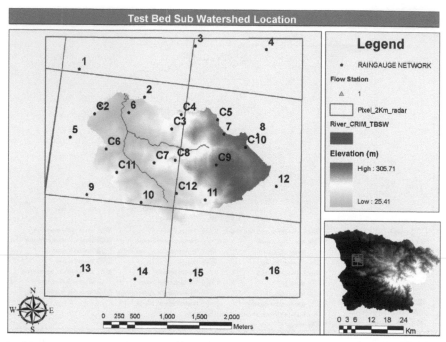

FIGURE 3.7 TBSW location within the 4 × 4 km² NEXRAD pixel and rain gauge network.

3.3 FLOW DIRECTION AND STREAM DEFINITION

For MBDB, the model comprises the Río Grande de Añasco, Río Guanajibo and Río Yaguez watersheds. Overland slope, flow direction, and stream locations were determined from the USGS 30 × 30 m² digital elevation model (DEM) quadrangles and resized to 200 m spatial resolution. During this step, the streams were "burned" into the model grid using a multistep process in ArcGIS, in which the flow direction is forced to follow the rivers. This step is necessary because the flow direction calculation tends not to be accurate in low slope areas (e.g., floodplains of the rivers). The final resized digital elevation model has correct flow direction based on the hydrological maps of the topographic quadrangles.

The flow direction and subsequent products were calculated with Arc Hydro Tools and ArcGIS 9.3. A flow direction map is necessary to calculate the flow accumulation map and create the stream network map. The flow accumulation is an accounting of cells contributing flow to a selected observation point, increasing the contributory area for observation points located further downstream. A cell located at the watershed outlet has the total cell number that drain to this point. The stream definition required 90 cells of flow accumulation to begin a channel. The river grid generated was used to define the channel cells in Vflo (Figure 3.8).

The TBSW model was developed using the same procedure described above but using the 10 m DEM [25]. The flow direction and stream definition were used to define the overland and channel cells respectively; based on the sub watershed delineation and river definition shown in Figure 3.7.

3.4 CHANNEL GEOMETRY

Channel geometry in the hydrologic model is necessary for the channel cells or cross section cells in the model and includes the sides slopes, cross sectional data or base width for trapezoidal assumption and channel slope. The geometry would affect the flow response, increasing the stages for narrow rivers and decreasing stages for wide rivers, principally due to the

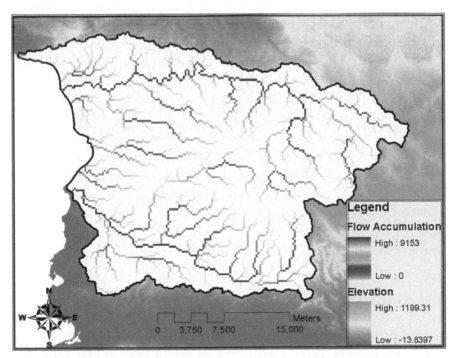

FIGURE 3.8 Flow accumulation and stream definition for Río Grande de Añasco, Río Guanajibo and Yagüez basin model.

storage. MBDB is not characterized by large width variations over short distances; typically widths are within the range of 3–5 m for upland rivers and creeks and up to 32 m for low lands according to measurement samples in aerial photos taken December, 2006 (Google Earth) over the study area.

The channel slide slopes were assumed to be 1:1 for the streams where no cross section information was available. The stream geometry was defined with data collected in 2002 by the PRWRERI [76, 135]. At Río Grande de Añasco, 25 cross sections were measured along the river; 10 cross sections were surveyed in Río Guanajibo, located downstream of PR-114 and in Río Yagüez only four cross sections were measured upstream of the channelized section. To define the flood plain within the cross sections, an extending process was made using the digital terrain model (10 m resolution) and creating interpolation lines to extract the entire cross section and new cross sections. Additional cross sections were extracted from DEM (10 m resolution) to characterize the flood plain where no field cross sections were surveyed and a simple trapezoidal river section was used measuring the river width from 2006 aerial photos of Google Earth, 2006 and the side slope set to 1:1. Figure 3.9 shows the locations of cross sections extracted from the DEM for the Río Guanajibo and Río Grande de Añasco. The channel slope was determined using the stream definition raster layer (Figure 3.8) and the slope map calculated with the DEM at 10 m resolution for the stream reaches where no survey data was available.

The stream map generated with the DEM at 10 m resolution was used to define the channel cells in *Vflo* for the TBSW model; channel side slopes were assumed to be 1:1; and bed channel width was set to 5 m. In most of the river sections (measured from Google Earth), the channel width is about 5 to 10 m, supposing bed width is about 4 to 8 m. Streamflow and flow volume are not sensitive to bed width; however, the stream stage is sensitive to bed width according to some tests realized.

3.5   STAGE AND RATING CURVE FOR THE TBSW CREEK

A pressure transducer was installed at the TBSW outlet to collect flow stage measurements every 5 min from October 20, 2007 to May 2009. The

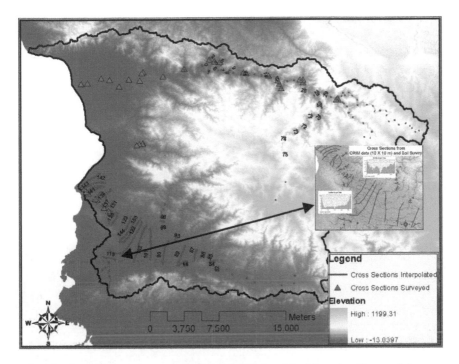

FIGURE 3.9   Cross sections surveyed and interpolated for Mayagüez Bay model.

instrument was located at 18.232667° latitude; −67.119533° longitude and elevation of 25 m amsl (see Figure 3.11). Daily minimum barometric pressures were used to correct the factory calibrated stage measurements using the Miradero KPRMAYAG1 weather station (18.2° north latitude, 67.13° west longitude and elevation of 22.86 meter above mean sea level), available at www.weatherunderground.com. The average adjusted stage value was calculated in 0.847 m with 0.0225 m standard deviation. This value was using the minimum pressure measured at Miradero KPRMAYAG1.

Stream cross-sections and bed slopes were measured in the field (Figure 3.10) and the rating curve was generated using HEC-RAS 4.0 hydraulic model [Hydrologic Engineering Center, 53] with 3 cross sections and slopes observed. The downstream boundary condition was assigned as critical depth and flows were assigned with subcritical flow condition. The full bank stream-rating curve was fitted to the following third order polynomial equation (Eq. 1) with a regression coefficient of 1,

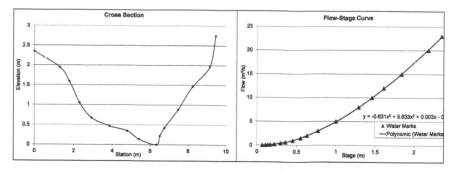

FIGURE 3.10   Cross section measured at the instrumentation place and rating curve to full bank condition.

FIGURE 3.11   Principal channel bed at TBSW (right) and location of the pressure transducer (left).

where flow is in cubic meters per second and stage in meters. The Eq. (1) was used to convert stage elevations to flow discharge for the events.

$$Flow = -0.631 \, stage^3 + 5.633 \, stage^2 + 0.003 \, stage - 0.0631 \quad (1)$$

To setup the distributed model at TBSW, information was assigned to selected model cells corresponding to the principal stream channel. The bed channel slopes for the TBSW model were assigned by segments using the average longitudinal slope between cross sections digitized from the DEM (10 m) and corroborated with field measurements. Figure 3.11 shows pictures of the outlet section and the pressure transducer location. The TBSW creek was divided in three creeks (Figure 3.12). The Lower Creek has a longitudinal average slope of 1.25% and Upper Creek has

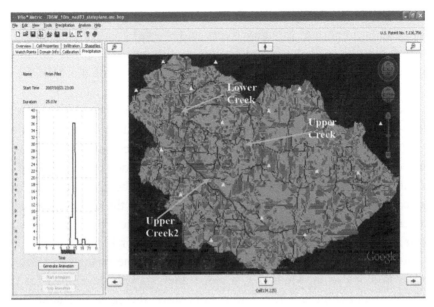

FIGURE 3.12 TBSW hydrologic model configuration (Vflo) and identification of the river reaches.

2.22%. Upper Creek 2 is shown in Figure 3.12 and was divided into two segments, the upstream segment shows a slope of 11.27% and the downstream segment is 3.27%. Figure 3.12 shows the *Vflo* model with the channel and overland cells at 10 m resolution and the locations of the creeks named above.

3.6 SLOPE ANALYSIS

Land surface slope is another important source of uncertainty in hydrologic modeling. High (low) slopes affect the time to peak producing early (retarded) peaks, less (more) infiltration, increasing (decreasing) discharge volume and increasing (decreasing) peaks. The average and standard deviation of the slope for Río Grande de Añasco basin were 34.6% and 21.7% respectively; for Río Guanajibo basin 28.2% and 22.4%, respectively; for Río Yagüez 29.8% and 18.0%, respectively; and for TBSW were 31.0% and 14.9%, respectively, calculated with the DEM at 10 m resolution.

Figure 3.13 and Table 3.3 show the subwatershed map and the average land surface slope values and standard deviation for each watershed and

subwatersheds for MBDB area. In total, 24 subwatersheds were identified
for the most important tributary rivers and coastal areas, the majority of
those exhibiting high slopes and similar conditions to the TBSW, indicating that the TBSW could be a representative sample of the MBDB, in
terms of the slope parameter.

Maintaining the land surface slope values when resampling techniques are used would improve the flow prediction at larger terrain scales.
A method to calculate slope at different grid size resolutions was investigated without decreasing of slope. Different methods can be applied to
calculate the resampled slope while the up scaling is being done. The slope
up-scaling was performed using two methods and three resample techniques for the TBSW model using ArcGIS 9.3. The TBSW presents an
average slope of 31.03% with a standard deviation of 14.93%.

To verify the results and obtain a box plot of the change and degradation in slope using Method 1, a slope analysis was developed for the
MBDB model (Figure 3.13). The results show the same degradation of
the mean slope (dashed lines: Figure 3.16) using Method 1 and the nearest

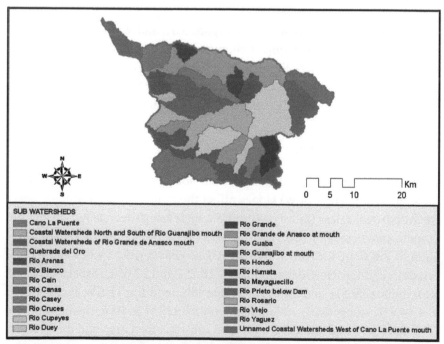

FIGURE 3.13   Sub Watersheds map belonging to MBDB.

TABLE 3.3 Mean Land Surface Slope and Standard Deviation for the Subwatersheds

Watershed Name	Sub Watershed Name	Area (km²)	Mean Slope (%)	Standard Deviation (%)
Río Grande de Añasco	Unnamed Coastal Watersheds West of Cano La Puente mouth	28.78	28.70	21.70
	Río Humata	12.65	35.75	17.79
	Cano La Puente	28.37	20.11	25.65
	Río Grande de Añasco at mouth	101.91	32.30	20.85
	Río Arenas	15.41	28.72	14.57
	Río Casey	29.64	37.11	18.87
	Río Blanco	31.45	44.09	20.17
	Coastal Watersheds of Río Grande de Añasco mouth	18.13	7.39	10.69
	Río Mayaguecillo	18.11	37.81	17.75
	Río Cañas	38.00	26.72	16.10
	Test Bed Sub-Watershed	**3.56**	**31.03**	**14.93**
	Río Guaba	83.20	46.06	19.38
	Río Prieto below Dam	43.31	41.51	18.43
	Total area and average slope	**448.95**	**34.60**	**21.67**
Río Yagüez	Quebrada del Oro	6.74	19.76	16.56
	Río Yagüez	35.24	31.67	17.69
	Total area and average slope	**41.98**	**29.76**	**18.05**
Río Guanajibo	Río Rosario	62.15	38.02	20.59
	Coastal Watersheds North and South of Río Guanajibo mouth	21.03	11.92	15.83
	Río Hondo	12.52	25.49	17.06
	Río Guanajibo at mouth	81.35	17.81	17.07
	Río Duey	35.70	37.25	19.06
	Río Cain	21.13	39.02	17.99
	Río Grande	25.41	47.21	23.64
	Río Cruces	19.55	38.39	22.83
	Río Cupeyes	11.03	39.55	19.14
	Río Viejo	60.65	15.71	18.37
	Total area and average slope	**350.52**	**28.17**	**22.38**

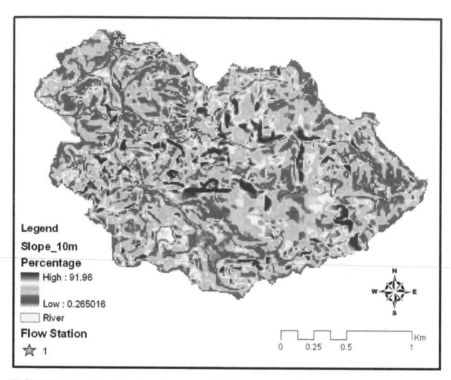

FIGURE 3.14   Land Surface slope map for the TBSW, slope values in percent.

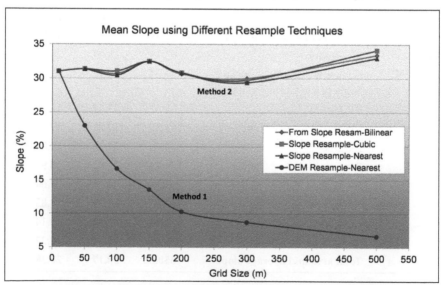

FIGURE 3.15   Slope calculated for TBSW using different resample techniques.

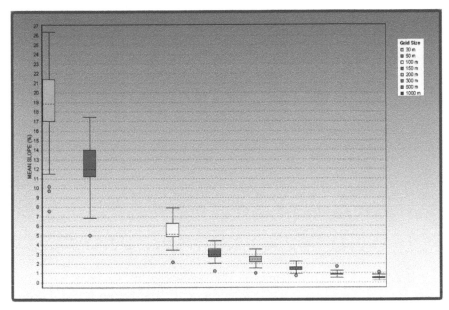

FIGURE 3.16 Slope box plots (quartiles 25 and 75) for the MBDB study area calculated with Method 1 and nearest neighbor resample technique, mean slope (dashed lines), quartiles 5 and 95 (solid lines) and outliers (dots).

neighbor resampling technique at DEM resolutions of 30, 50, 100, 150, 200, 300, 500 and 1000 m.

Figure 3.16 presents slope degradation in terms of the interquartile 25–95 (solid boxes), interquartile 5–95 (solid lines) and outliers (dots). Figures 3.17 and 3.18 present spatial graphical representation of the slope degradation using the two methods described above. The same interval classes were chosen to represent the slope. Method 2 in Figure 3.18 presents much more area in red color than Method 1 in Figure 3.17, because it presents more areas without degradation and slope values greater than 16%. Therefore, Method 2 is the recommended for up-scaling both the slope of TBSW and Mayagüez Basin model.

3.7 GREEN-AMPT INFILTRATION MODEL: PARAMETERS ASSIGNMENT

The abstractions in the distributed hydrologic model are calculated with the Green-Ampt infiltration model. The principal parameters are: saturated

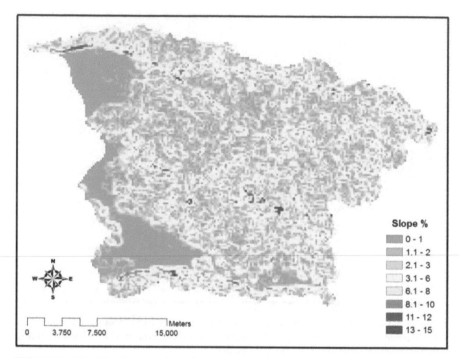

FIGURE 3.17   Visual comparisons between resample methods at 200 m resolution for the MBDB model by Method 1.

hydraulic conductivity; effective porosity, soil depth, and wetting front. Parameter values were assigned using the SSURGO maps and database from the USDA [106–109], which contains the soil classes for Puerto Rico. Initially, the soil map was classified into six basic textures and the hydraulic conductivity, wetting front and effective porosity values were assigned from literature as shown in Table 3.4 [9, 34, 65, 128]. Using the Book Reference values of infiltration parameters from Table 3.4, average parameter values were calculated for the tributary area at the streamflow gauge stations, located in the watersheds. Average parameter values in several flow meter stations are indicated in Table 4.

At Río Grande de Añasco near San Sebastian for example, the average hydraulic conductivity is 0.05 cm/h, the wetting front is 28.29 cm, the effective porosity is 0.364, and the soil depth assigned uniformly to the basin area was 20 cm. A preliminary study was developed with the infiltration values shown in in Table 3.4.

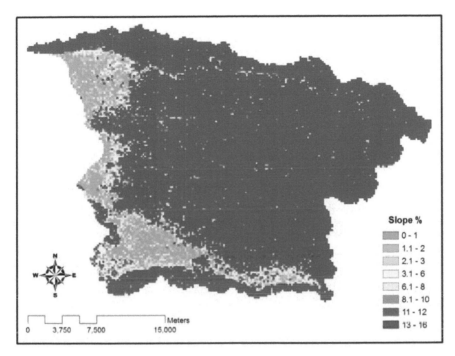

FIGURE 3.18 Visual comparisons between resample methods at 200 m resolution for the MBDB model by Method 2.

The volume calculated was over predicted in almost all cases. Therefore an exhaustive analysis was conducted to enhance the infiltration parameter values since the literature shows low hydraulic conductivity values using the texture class approach. In Puerto Rico, the soils present high organic matter content and some clays are well drained, and are considered as hydrologic group B, for example Alonso, Consumo, Delicias and Maricao soils [SSURGO]. New values for hydraulic conductivity, total porosity and effective porosity were obtained using the percentage of sand, silt and clay and average bulk density from the SSURGO database and *Rosseta Lite program* [86–90] from HYDRUS-1D [96]. Rosetta implements pedotransfer functions to predict van Genuchten [114] water retention parameters and saturated hydraulic conductivity (K_s) by using textural class, textural distribution, bulk density and one or two water retention points as input. Rosetta follows a hierarchical approach to estimate water retention and K_s values using limited or more extended sets

TABLE 3.4　Summary of the Infiltration Values for the Green Ampt Model

Basin	Soil texture	Effective porosity	Wetting front (cm)	Hydraulic conductivity (cm/h)
Book reference	Sand	0.42	4.95	11.78
	Loam	0.43	8.89	0.34
	Clay Loam	0.31	20.88	0.10
	Clay	0.39	31.63	0.03
	Gravel	0.24	1.5	2.27
	Rock	0.17	1	0.036
Average Values over the Watersheds				
Añasco near San Sebastian	—	0.364	28.29	0.05
Guanajibo near Hormigueros	—	0.33	22.5	0.1
Río Rosario	—	0.328	25.2	0.03
TBSW	—	0382	31.21	0.03
Río Casey	—	0.376	30.41	0.03
New Average Infiltration Values				
Añasco near San Sebastian	—	0.412	28.61	0.75
Guanajibo near Hormigueros	—	0.363	22.85	6.35
Río Rosario	—			
TBSW	—	0.43	31.57	0.69
Río Casey	—	0.418	30.41	0.64

of input data [87–89]. The calibration data for Rosetta has a set of 2134 samples for water retention and 1,306 samples for K_s [88] distributed in USA and some from Europe. The authors suggested that the usage of Rosetta for other climate zones, and hence other pedogenic processes, could lead to inaccurate predictions.

3.7.1　ASSUMPTIONS FOR UNCLASSIFIED SOIL CLASSES

Some soils did not have bulk density and percentage of sand, silt and clay. In these cases assumptions were made for alluvial land, leveled clayed classification, limestone, gravel, pits and quarries, serpentine rock, volcanic rock and limestone rock as described in this section.

3.7.1.1 Alluvial Land

Alluvial land has a variable profile, is a fine-grained fertile soil deposited by water flowing over flood plains or in river beds. Clay or silt or gravel are carried by rushing streams and deposited where the stream slows down. The Soil Conservation Service classified this soil in the hydrologic group D and reports that the alluvial land has 0–1 inches of ponding depth range, very long ponding duration and floods frequently during the year [USDA, 106–109]. Therefore, it is assigned a classification of Clay with an effective porosity of 0.475, 31.63 cm suction head and 0.06 cm/h saturated hydraulic conductivity.

3.7.1.2 Leveled Clayed

Leveled Clayed presents a hydrologic group C. The hydraulic conductivity value assigned to this classification was the average value between clay texture and hydrologic group C and it was 1.225 cm/h with a range between 0.801 and 2.789 cm/h. The same procedure as was used for alluvial land was used for leveled clay where the effective porosity was assigned the average value of 0.427 and a value of 31.63 cm for suction head, as recommended for clay.

3.7.1.3 Limestone

Limestone is a sedimentary rock composed largely of the mineral calcite (calcium carbonate: $CaCO_3$). The hydraulic conductivity was 570 cm/h, taken from Freeze and Cherry (1979), the range for this value varies from 0.11 to 1,142 cm/h. The effective porosity is 0.14. The wetting front suction head was set to 1 centimeter, the minimum for sand reported by Vieux [126].

3.7.1.4 Gravel, Pits and Quarries

Gavel, pits and quarries have a hydrologic group A, assigned in SSURGO database [USDA, 106–109] meaning that they possess very good infiltration. The values assumed for their classification was medium gravel with a moderate degree of sorting and without silt content. For this material, the saturated hydraulic conductivity was assigned a value of 297 cm/h and an effective porosity of 0.24. The wetting front suction head was the minimum for sand reported by Vieux [127] of 1 cm.

3.7.1.5 Serpentine Rock

According to Freeze and Cherry [34], the saturated hydraulic conductivity (Ks) for fractured metamorphic and igneous rocks is between 0.00114 and 11.4 cm/h, the average is 5.71 cm/h. The effective porosity assigned was 0.26 for metamorphic rock.

3.7.1.6 Volcanic Rock

Volcanic rocks are usually fine-grained or aphanitic to glassy in texture and are named according to both their chemical composition and texture. Basalt is a very common volcanic rock with low silica content. For Basalt rock we assumed a total 0.17 (reported range of 0.03 to 0.35); effective porosity 0.1 and saturated hydraulic conductivity 570 cm/h for fractured basalt (10 to 10^5 m/year).

The values assigned to *Soil not Surveyed* classification were average hydraulic conductivity for clay texture in the whole study area: 1 cm/h; and the effective porosity and wetting front suction values correspond to clay as reported by Vieux [127]. For the TBSW model, all the parameters were assigned to a grid model resolution of 10 m from the MBDB model. Average infiltration parameters for the TBSW are tabulated in Table 3.5 with detailed soil names and parameter values used. Bouwer [19] suggested multiplying the hydraulic conductivity by 0.5 for the saturated hydraulic conductivity in Green-Ampt model. Therefore the average saturated hydraulic conductivity for the TBSW is 0.69 cm/h.

3.8 SOIL DEPTH

The soil depth is a very important parameter to calculate the infiltration losses. The USDA [106–109] reports the soil depth for each soil when some restrictive layer or lithic rocks exist at a shallow depth. In other cases a maximum soil depth is assigned a value of 152 cm (60 inches), corresponding to the depth surveyed. Lithic is a continuous hard rock and less permeable, in some cases it is encountered at a depth of 10 cm from the soil surface. For some soils a paralithic rock is present under the layered soil. The paralithic rock is a weathered layer and broken rock in contact with fissures less than 10 cm apart, which allow roots and water to penetrate the underlying rock. Major hydraulic conductivity is allowed, and works like fractured rock. Soils under this condition are allowed to increase the soil depth to 600 cm indicating no depth restriction, and other soils without any restrictive layer or lithic rock were set to 300 cm, almost double that of the survey. In this way the soil depth assigned to the soil map will be

TABLE 3.5 Soil Classification (SSURGO), Hydrologic Group and Infiltration Parameters at TBSW

Soil Name	Texture	Hydrologic Group	Area (%)	Wetting front (cm)	K_s (cm/h)	Depth (cm)	Effective porosity
Consumo	Clay	B	59.85	31.63	1.273	300	0.415
Dagüey	Clay	C	15.11	31.63	1.266	300	0.451
Humatas	Clay	C	25.03	31.63	1.736	300	0.454
Serpentinite	Rock Serpentine	D	0.01	3.00	5.7	300	0.26
Toa	Silty Clay Loam	B	0.01	27.30	0.294	300	0.377
Average	—	—	—	**31.62**	**1.38**	—	**0.43**

TABLE 3.6 Resized Grid Area for the Land Use Map [75]

Re-class name	Manning roughness (n)	Impervious (%)	Area with 30 m (km²)	Area with 200 m (km²)	Δ Area (km²)
Agriculture	0.166	5	54.93	55.92	0.99
Agriculture/hay	0.190	4	0.13	0.12	–0.01
Forest, shrub, woodland and shade coffee	0.191	2	529.16	529.12	–0.04
Other emergent wetlands	0.050	1	1.26	1.24	–0.02
Pasture	0.225	5	172.84	173.2	0.36
Quarries, sand and rock	0.020	95	0.75	0.56	–0.19
Urban and barren	0.080	81	60.02	58.68	–1.33

the maximum possible and reductions would be considered for calibration proposes. Values assigned for the TBSW area are shown in Table 3.5.

3.9   ASSIGNING OVERLAND ROUGHNESS, IMPERVIOUS AND CROP COEFFICIENT

Overland roughness is an input parameter in hydrologic models and this parameter affects principally the peak flow in a hydrograph. Two sources were analyzed to determine the land use in the area. One source was obtained from land use/land cover map for Puerto Rico [51], which was reclassified by Prieto [75] into six land use classes. Appropriate Manning's and impervious values were assigned to each class at 30 m resolution (Table 3.6). A resize from 30 m to 200 m will change the area distribution of some land use and would affect the flow response (e.g., flow volume). The land class most affected by resizing is the urban area showing a decrease in area of 1.33 km^2, followed by an increase in Agriculture by a 0.99 km^2, areas of special interest in terms of flooding (Table 3.6).

The sum of the land use map areas between 30 m and 200 m are different due to pixel sizes; 200 m is rougher and covers more area, while the 30 m pixel can adjust much better to the basin form.

The second land use source was from remote sensing classification and field verification from PRWRERI [76] shown in Figure 3.6 with 35 classes. The land use classification was reclassified into 13 classes and is shown in Figure 3.21. The roughness values were specified for each class according to literature and expertise and shown in Table 3.6. A value of 0.118 is the average roughness value for the MBDB model and 0.12 for the TBSW.

Another parameter that is contingent upon the land use classification is the crop coefficient. Its coverage was determined using the land use classes derived in Figure 3.19 at 30 m resolution. Values of K_c (mid-season crop stage) were assigned from Allen et al. [4] and are shown in Table 3.7. Allen et al. [4] did not present K_c values for forest land use. Therefore, an apple tree with active ground cover class value was assumed (for possible representation of forest), with a maximum of 1.2 K_c. The TBSW exhibits a predominant forest land use (see Figure 3.20, 30 m resolution) of low density with 39.36% of the area; brush rangeland with 38.17% of the area and 14.51% urban land use, respectively (Table 3.8). The Figure 3.21 shows some images taken for the forest representation and urban area.

TABLE 3.7   Land Use Classification with the Manning Roughness Values and Crop Coefficient (K_c) for MBDB

Classes	Re-classification	Manning roughness (n)	K_c	Area (m²)
Coffee	Agricultural Land	0.080	1.100	15.76
Coffee, orange		0.080	1.000	0.01
Coffee, orange, plantain/banana		0.080	1.000	0.01
Coffee, plantain/banana		0.080	1.100	12.73
Coffee, plantain/banana, oranges		0.080	1.025	0.33
Dairy Farm or dairy cow feeding		0.050	0.400	0.03
General agriculture		0.080	1.000	1.17
Nurseries and ornamental horticulture		0.080	1.000	0.39
Orange		0.080	0.850	0.66
Orange, coffee		0.080	0.950	0.64
Orange, plantain/banana		0.080	0.900	0.29
Orange, plantain/banana, coffee		0.080	1.000	0.02
Plantain/banana		0.080	1.200	7.21
Plantain/banana, coffee		0.080	1.150	0.06
Plantain/banana, coffee, oranges		0.080	1.200	0.49
Plantain/banana, orange		0.080	1.025	0.11
Shade coffee plantation		0.080	1.100	0.06
SUB-TOTAL		**0.078**	**0.992**	**39.99**
Barren land	Barren Land	0.015	0.300	10.18
Forest high density	Forest high density	0.150	1.200	156.19
Forest low density	Forest low density	0.150	1.100	234.31
Forested Wetland	Forested Wetland	0.070	1.200	2.83
Native pastures	Native pastures	0.045	0.850	6.73
Non-Forested Wetland	Non-Forested Wetland	0.050	1.100	2.16
Pasture	Pasture	0.035	0.950	1.50
Shrub and brush rangeland	Range Land	0.130	1.000	248.92

TABLE 3.7   Continued

Classes	Re-classification	Manning roughness (n)	K_c	Area (m²)
Bare exposed rock	Rocks	0.015	0.100	0.04
Gravel pit		0.015	0.100	2.07
Transition area	Transition area	0.050	0.300	0.79
Transportation, communication	Urban or Built-Up	0.015	0.300	11.78
Urban or built-up land		0.015	0.300	97.40
Waste disposal areas		0.015	0.300	0.44
Pond	Water	0.030	1.050	0.24
Streams and canals		0.030	1.050	2.97
TOTAL		**0.188**	**0.966**	**818.53**

Source: PRWRERI [76] for classes and Allen et al. [4] for K_c.

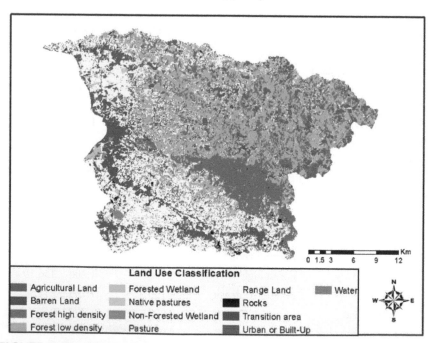

FIGURE 3.19 Land Use general reclassification from Land Sat[ET] 2004, *Source:* PRWRERI, (2004). Land Use Classification of the Mayagüez Bay Watershed (Río Grande de Añasco, Río Yagüez, and Río Guanajibo Watersheds. Puerto Rico Water Resources and Environmental Research Institute (PRWRERI). Developed for the Puerto Rico Environmental Quality Board.

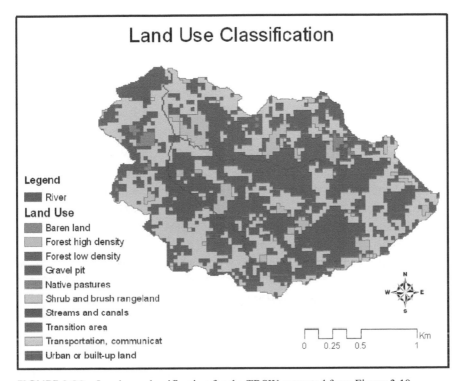

FIGURE 3.20 Land use classification for the TBSW extracted from Figure 3.19.

FIGURE 3.21 The land use of the TBSW.

TABLE 3.8 Land Use Classification, Manning Roughness (n) Values and K_c for Evapotranspiration Quantification in the TBSW

Land use classification	Manning roughness (n)	K_c	Area (km²)	Area %
Barren land	0.0150	0.300	0.0378	1.06
Forest high density	0.1500	1.200	0.2083	5.86
Forest low density	0.1500	1.100	1.3994	39.36
Gravel pit	0.0150	0.100	0.0018	0.05
Native pastures	0.0450	0.850	0.0009	0.03
Shrub and brush rangeland	0.1300	1.000	1.3570	38.17
Streams and canals	0.0300	1.050	0.0045	0.13
Transition area	0.0500	0.300	0.0216	0.61
Transportation, communication	0.0150	0.300	0.0083	0.23
Urban or built-up land	0.0150	0.300	0.5157	14.51

3.10 EVAPOTRANSPIRATION

The hydrologic model requires potential or reference evapotranspiration as input to dry the soil in a long-term simulation. This section identifies the uncertainties associated with the evapotranspiration quantification, because this parameter is time and scale dependent and is related to the meteorological stations located within the area of interest. Reference evapotranspiration can be calculated by the Penman-Monteith method (Eq. 7) and the Hargreaves Samani method (Eq. 8) using data from the NRCS Soil Climate Analysis Network (SCAN) weather stations located in western and southern Puerto Rico. Two stations are located within the MBDB and relatively close to the TBSW (i.e., the USDA Tropical Agricultural Research Station (TARS) at Mayagüez and Maricao Forest, PR). Penman-Monteith [4] and Hargreaves-Samani [45] methods were compared at the stations mentioned with a daily time step from October, 2007 to October 2009. The FAO56 Penman Monteith evaporation equation is presented below [4]:

$$ET_O = \frac{0.408 \cdot \Delta \cdot (R_n - G) + \gamma \cdot \left(\dfrac{900}{T + 273}\right) \cdot u_2 \cdot (e_s - e_a)}{\Delta + \gamma \cdot (1 + 0.34 \cdot u_2)} \qquad (2)$$

where, ET_0 is reference evapotranspiration (mm/day), Δ is slope of the vapor pressure curve (kPa/°C), R_n is net radiation (MJ/m²day), G is soil heat flux density (MJ/m²day), γ is psychometric constant (kPa/°C), T is mean daily air temperature at 2 m height (°C), u_2 is wind speed at 2 m height (m/s), e_s is the saturated vapor pressure and e_a is the actual vapor pressure (kPa).

Equation (7) applies specifically to a hypothetical reference crop with an assumed crop height of 0.12 m, a fixed surface resistance of 70 sec/m and an albedo of 0.23. The Hargreaves-Samani equation for reference or potential evapotranspiration [45] is given below:

$$PET = 0.0135 \times R_s \times (T_{ave} + 17.8) \tag{3}$$

where, R_s is solar radiation in units of mm/day and T_{ave} is average air temperature (°C). R_s is readily converted from units of MJ/m²day to equivalent depth of water in mm/day by dividing by the latent heat of vaporization (2.45 MJ/m²day).

The Pearson correlation coefficient (R^2) between Eqs. (7) and (8) was 0.9375 and the bias was 0.956 for this period, indicating that the Hargreaves Samani constant (0.0135) presented in Eq. (8) could be corrected by a factor of 0.956 for the current study area using a more simplistic formula than FAO-Penman-Monteith equation (Eq. 7). Goyal et al. [40] developed monthly linear regression equations for air temperature (mean temperature (T_{ave}), maximum temperature (T_{max}) and minimum temperature (T_{min}) for Puerto Rico, which depend on the surface elevation (m). PET can be calculated using these linear regressions [44] and Hargreaves-Samani equation [45] extended for places where no solar radiation data is measured.

$$PET = 0.0023 \times R_s \times (T_{ave} + 17.8)(T_{max} - T_{min})^{0.5} \tag{4}$$

where, PET is potential or reference evapotranspiration (mm/day) and R_a is the extraterrestrial radiation (mm/day).

Solar radiation is highly spatially variable in Puerto Rico [48, 49]. Therefore, the effectiveness of Eqs. (3) and (4) to estimate PET using the temperature versus elevation relationships developed by Goyal at short time scales (daily) was evaluated in the current study. Constants in Goyal's monthly linear regressions were interpolated to daily constants [40]. All input parameters needed in the Hargreaves-Samani methods (Eqs. 3 and 4) are measured by the SCAN stations.

The elevation at the TARS is 13.72 m amsl with an average temperature (T_{ave}) of 23.9°C for the period of analysis (October, 2007 to October, 2009); and in Maricao Forest the elevation is 747 m with T_{ave} 19.7°C. The results show that the Goyal regressions at a daily time step predict the T_{ave} with a coefficient of determination R^2 of 0.46 for TARS and 0.62 for Maricao [40]. However, if PET is calculated with the solar radiation measured at the stations along with the T_{ave} derived from the Goyal regressions [40], the improved R^2 of 0.987 and 0.992 are obtained at TARS (Figure 3.22) and Maricao Forest (Figure 3.23), respectively.

Values of R^2 of 0.2145 for TARS and 0.0013 for Maricao were obtained using Goyal's elevation model [40] and Eq. (4). The R^2 is increased to 0.2254 for the Maricao station if the PET is calculated using the T_{ave} from the equations by Goyal [40] and the solar radiation is assumed to be equal to the TARS solar radiation (Figure 3.22).

These results show that solar radiation is a spatially sensitive parameter in the PET calculation and that solar radiation cannot be assumed equal at locations distant from each other. Remotely sensed satellite measurements are suggested for a better spatially distributed solar radiation dataset, according to Harmsen et al. [48, 49]. For a long-term hydrologic model, simulations for the TBSW, we used the PET calculated using Eq. (3) and assuming that the solar radiation is the same as TARS, due to its relatively close proximity to the TBSW, around 2.5 km, compared to 16.3 km between the TBSW and Maricao Forest stations. Although not used in this study, another option would have been to use the daily operational solar radiation data described by Harmsen et al. [48] for Puerto Rico [http:/ pragwater.com/solar-radiation-data-for-pr-dr-and-haiti/].

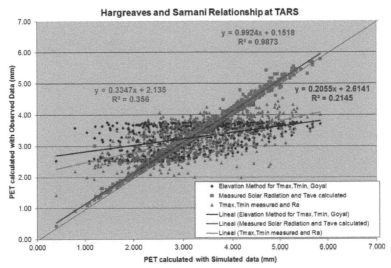

FIGURE 3.22 Potential Evapotranspiration with Hargreaves-Samani relationship for observed T_{max}, T_{min}, T_{ave}, solar radiation, extraterrestrial radiation; and temperatures predicted by Goyal relationships at TARS station. *Source:* Goyal, M. R., E.A. González and C. Chao de Báez, (1988). Temperature versus elevation relationships for Puerto Rico. J. *Agric. UPR72*(3), 449–67.

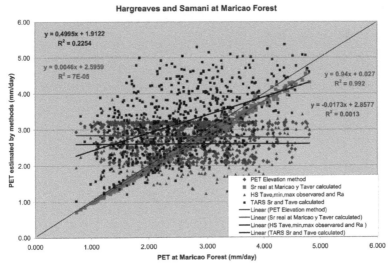

FIGURE 3.23 Potential Evapotranspiration with Hargreaves-Samani relationship for observed T_{max}, T_{min}, T_{ave}, solar radiation, and extraterrestrial radiation; and temperatures predicted by Goyal relationships at Maricao Forest station. *Source:* Goyal, M. R., E.A. González and C. Chao de Báez, (1988). Temperature versus elevation relationships for Puerto Rico. J. *Agric. UPR72*(3), 449-67

CHAPTER 4

FLOOD PREDICTION LIMITATIONS IN SMALL WATERSHEDS: METHODOLOGY [1,2]

ALEJANDRA M. ROJAS-GONZÁLEZ

CONTENTS

[1] This chapter is an edited version from, "*Alejandra María Rojas González, 2012. Flood prediction limitations in small watersheds with mountainous terrain and high rainfall variability. Unpublished PhD Thesis for Department of Civil Engineering and Surveying, University of Puerto Rico – Mayagüez Campus*".

[2] Numbers in brackets refer to the references at the end of this book.

4.1   INTRODUCTION

This chapter presents the technical methodologies used in this research to address the research questions presented in Chapter 1. A determination of parameter sensitivity in the MBDB model is presented, where various parameters were first perturbed by multiplication factors to generate spider plots, and then the factors 0.5 and 1.5 (representing ±50%) were used to calculate the relative sensitivity (Sr) for different variables and events. Using the TBSW model, some parameter aggregation techniques are evaluated for later use in the up-scaling experiment. This section presents the evaluation of uncertainties in Quantitative Precipitation estimates from MPE by comparison with a high density rain gauge network; and a methodology to evaluate uncertainty due to hydrologic model (grid spacing) and rainfall resolution were addressed.

To establish a flood alarm system in the MBDB, first, one must know the likelihood and uncertainty associated with a prediction due to the inputs and parameters variations. Some initial sensitivity tests were developed in the Mayagüez Bay model to understand how some parameters and inputs affect the flow prediction. The major sources of uncertainties are associated with inputs such as rainfall estimation, terrain slope, parameter values and initial conditions; and all these sources of uncertainty are resolution-dependent. How much rainfall variation is there at scales below the radar pixel size and how much does rainfall variation and DEM resolution affect predictability? These questions will be addressed in the TBSW analysis.

The TBSW is useful for research purposes and represents a "real world" laboratory to study the predictability limits due to aggregation of high-resolution inputs in a hydrologic model. In the TBSW (Figure 3.7 in Chapter 3 of Part I), a dense rain gauge network was installed as part of this investigation and a pressure transducer for water level measurements. Other high-resolution data exists for the TBSW including topography [digital elevation model, 25]; soils and land use maps, etc. These sets of information are ideal to define how much detail is necessary in the physical modeling process and the value of increasing the rainfall resolution, as well as the hydrologic model grid resolution within small watersheds. Carpenter [23] mentioned that the uncertainty in the model output is inversely proportional to the watershed area. In other words, for a small hydrologic model, a large degree of uncertain exists at the subwatershed scale. Therefore, the magnitude and behavioral impact of the rainfall errors in the hydrologic forecasts help to define the precision and accuracy necessary in new rainfall algorithms and radar technologies. New radar technologies are being developed under the CASA project at UPRM [22] and are available for western Puerto Rico, promising higher resolution than NEXRAD, and will be a critical component in the flood alarm system.

Evaluating possible CASA radar resolution in this study with the rain gauges information, authors of this study determined the predictability and quantify the uncertainty due to terrain and rainfall grid size resolution at scales below the typical radar resolution (2×2 km^2 cell size) in small subwatersheds. After finding the predictability limits and assessing the predictability in the TBSW, they formulated recommendations to initialize the larger model (MBDB) and enhance the flood prediction in mountainous basins. All statistical analyzes in this research were performed using Minitab 16 [67].

The following sections describe the methodology and activities required to achieve a successful investigation and to address the research questions presented before in Chapter 1. For convenience, a summary of the research questions are listed here:

- How flood prediction is affected by the spatial variability of point rainfall at scales below that of the typical resolution of radar-based products?
- How does the DEM and parameter aggregation affect the model's predictive capabilities and the errors of the hydrologic system?
- Would the assumptions developed for the small scale enhance the hydrologic predictability at larger scales?

4.2 ADDITIONAL FIELD MEASUREMENTS

A dense network of rain gauges (28 tipping bucket rain gauges with data loggers) were installed within a single GOES Satellite Hydro-Estimator (HE) pixel (4×4 km^2) and 64% of the rain gauges are within TBSW with the objective to obtain high-resolution rainfall within the area. Complete records were collected since June, 2007 when the last 12 rain gauges were installed within the TBSW [47] with a temporal resolution of 5 min. The Euclidian distance was calculated between rain gauges within the TBSW, exhibiting a maximum range distance of 563.2 m and the mean distance was 218 m with a standard deviation of 99.5 m. The calculated mean Euclidian distance within the Hydro-Estimator pixel was estimated to be 334 m with a standard deviation of 171 m. The Figure 3.7 in Chapter 3 showed the location of the rain gauges network within the Hydro-Estimator pixel. Figure 4.1 shows the rain gauge network, the TBSW outline and the distance between rain gauges.

Additionally, a pressure transducer was installed at the TBSW outlet, which measured stage elevation data since October 2007 to May 2009 at 5 min temporal resolution.

4.3 EVALUATION OF PARAMETER AGGREGATION TECHNIQUES WITHIN THE TBSW

To develop the up-scaling experiment or set up any hydrologic model, it is necessary to evaluate which methodology is being addressed to create the

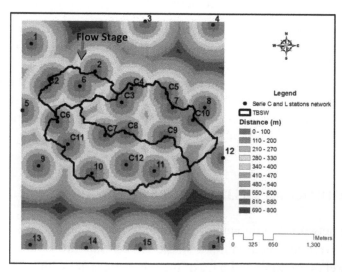

FIGURE 4.1   Rain gauge distribution and location within the HE pixel; TBSW location and Euclidean Distance between the stations.

hydrologic models at different resolutions. Several aggregation techniques are used in GIS to develop the parameters up-scaling. The aggregation consists of using data from the cells that will fall within the larger up-scaled cells and then applying to them mathematical operations to calculate a new aggregated cell value. All these aggregation techniques produce different results, which can affect the hydrologic response. Also, the order in which the slope is generated can alter the results. Two different orders were developed using different techniques and they are listed below:

1. Aggregate the terrain to a new resolution and calculate the slope for this resolution; or
2. Calculate the slope from high-resolution terrain model and then aggregate it to a new resolution.

The aggregation techniques and the order to derive slope were tested in the TBSW using Arc GIS tools. The tested resolutions were 10, 50, 100, 175, 250, 500 m, which generated graphs of how the slope has been degraded. A decision was taken as to which aggregation technique is best for the purposes of this research. Additionally the methodology was tested to see the degradation slope degree in the MBDB Model.

4.4 DETERMINATION OF HYDROLOGIC MODEL SENSITIVITY DUE TO PARAMETERS AND RAINFALL PERTURBATIONS FOR THE MBDB MODEL

To develop a distributed hydrologic model it is necessary to create an ensemble of different layers that represent the physical characteristics of the basin. Uncertainties associated with the model parameter values and their scales can be quantified by evaluating the hydrologic response given a range of parameter and rainfall perturbations.

The objective of this evaluation was to determine which parameters and rainfall are most sensitive in the mountainous areas, of the physical conditions present in Western Puerto Rico. Then these parameters were evaluated in the up scaling analysis. For this purpose, authors used the MBDB model at 200 m by 200 m cell resolution with three outlet points, summarizing different watershed characteristics in terms of area, shape and slopes.

The sensitivity analysis considered parameter and input perturbations by changing the magnitude of the parameter value, but not its spatial distribution. The multiplicative factors used to perturb the model and input (rainfall) were 0.5, 1.0, 1.5 and 2.0. The parameters used in the analysis were: overland and channel Manning roughness coefficient, the overland and channel saturated hydraulic conductivity, soil depth, and initial fraction of soil saturation. By demonstration in other studies, hydrologic models have been found to be sensitive to these parameters [126]. In this study, for completeness, we additionally evaluated the model response to variations in land slope.

Three important events that produced flash flooding in Puerto Rico were evaluated. The most important event with a recurrence greater than 100-year return period for Río Grande de Añasco River was Hurricane Georges in September 21–23, 1998. FEMA [32] estimated 4,078 *cms* at Río Grande de Añasco near San Sebastian for 100-year return period and the measured event had a peak of 4,587 *cms*. Other important events analyzed were November 11–16, 2003; and the Tropical Storm Jeanne on September 14–17, 2004. Interpolations of the rainfall amounts each time step (15 min) using the USGS rainfall stations available for each event in the MBDB area were made to obtain a distributed rainfall over the basins. The interpolation method used was the Exponential Weighted method.

The parameter and rainfall perturbations were evaluated at three basin outlets, which are: USGS 50144000 Río Grande de Añasco near San Sebastian, USGS 50136400 Río Rosario near Hormigueros and USGS 50138000 Río Guanajibo near Hormigueros.

Spider plots were used to evaluate the model response to the entire range of the parameters and to determine if there is a portion of the parameter range that yields unrealistic results. Spider plots for runoff depth and peak flow show the percent change in model output variable versus parameter value change (perturbation) by a given factor.

The Relative Sensitivity Coefficient (*Sr*) is defined as the ratio of the difference in the model output to the value of the output when the input parameters are set to their base values, divided by the ratio of change in the input parameter to the initial value of the input parameter as shown in Eq. (1):

$$Sr = \frac{\dfrac{\left(O_{P+\Delta P} - O_{P-\Delta P}\right)}{O}}{\dfrac{2\Delta P}{P}} \qquad (1)$$

where, O is model output with input parameters set at base values, P is the value of the input parameter, are model outputs with the input parameter plus or minus a specified perturbation (in this case ±50%).

The behavior of the relative sensitivity coefficient was evaluated using two variables: discharge volume in millimeters and peak discharge in cubic meters per second.

4.5 EVALUATION OF CURRENT QUANTITATIVE PRECIPITATION ESTIMATES

The NEXRAD radar is located near the City of Caycy at 860 m mean sea level and approximately at 120–130 km from Mayagüez city. It has been operational since 1999. Some errors exist associated with radar measurements due to factors such as distance from radar to the study area; the coverage gap between the terrain and radar beam (at western flood plains with a radar beam of 0.5 degrees a coverage gap between 1.8 and 2 km was found); and Z-R relationship applied. Mountain blockage at lower beam angles (0.35 to 0.45 degrees) affects the reflectivity received from some locations within the Añasco and Mayagüez flood plains. Figure 4.2 shows the detail of mountain blockage at beam angle of 0.35 degrees; for 0.5 degrees and higher blockage does not occur.

The NEXRAD radar resolution gives a spatial rainfall variability that fills the gaps between the rain gauges enhancing the spatial rainfall quantification. However, it is necessary to remove some bias between radar and rain gauges due to radar errors and rain rate quantification. Nevertheless, one may not know the rainfall variations at scales below the actual radar products (2×2 km² or 4×4 km²), because rain gauge networks do not exist at these scales within the island.

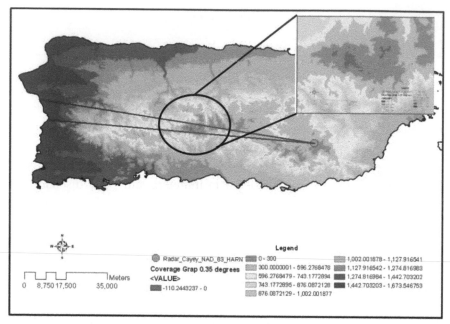

FIGURE 4.2   Coverage gap between terrain elevation and radar bean of 0.35 degrees with the detail of blockage at mountainous area.

4.5.1 EVALUATING RAINFALL DETECTION ACCURACY AND LONG-TERM BIAS QUANTIFICATION

Obtaining a long-term bias quantification between the radar and rain gauge network is an essential part of the uncertainty quantification. It is possible to observe and quantify how much change in the bias has occurred in time and magnitude. An evaluation of the MPE rainfall product and bias performance at hourly and daily temporal scales is evaluated within the Hydro-Estimator pixel for the year 2007 using the rain gauge network located in western Puerto Rico near the University of Puerto Rico – Mayagüez Campus, where the TBSW is located. Some rain gauges were not operating during some periods owing to gauge damage or low logger batteries, these data were eliminated from the analysis. Five-minute rain gauge data was accumulated to 1-hour and 1-day intervals, with the intention of comparing data with the original MPE temporal resolution and daily accumulations.

MPE pixels are based on a HAP (Hydrologic Rainfall Analysis Project) grid projection. Therefore, a geographic coordinate transformation from Stereographic North Pole to NAD 1983 State Plane Puerto Rico and Virgin Islands was performed for each hour using the ArcGIS project raster tool. The resampling technique algorithm used was the nearest neighbor assignment at 4×4 km^2 resolution.

The N1P rainfall product is calculated from NEXRAD as a rainfall rate every 5 or 6 min when the radar detects rainfall, and a 10 min N1P product is archived when no rainfall is detected. The N1P NEXRAD product originally has a polar geographic coordinate system (GCS) and using the NOAA Weather and Climate Toolkit program (NOAA National Climatic Data Center available at http://www.ncdc.noaa.gov) it is possible to transform the coordinates to GCS_WGS_1984. Different formats are available to export the data. The GIS shapefiles maintain the original orientation; however, in a distributed hydrologic model it is necessary to use raster or ASCII files to represent the spatial rainfall variation in the model. Due to raster characteristics it is not possible to maintain the original orientation.

The study was conducted with the projected and raster pixels, with the aforementioned in mind, 4 MPE pixels were obtained around the HE pixel. Area weights were calculated for intersecting areas between the MPE pixels and the HE pixel which are 0.281, 0.344, 0.169 and 0.206, respectively. These area weights are used to calculate an average map precipitation for each time step. Weights for the N1P radar product were also estimated for 9 partial N1P pixels within the HE pixel.

Long-term continuous validation between sensor rainfall estimates and rain gauge observations should be evaluated. The accuracy of rainfall estimates can be measured by decomposing the rainfall process into sequences of discrete and continuous random variables [78, 79, 142].

The discrete variables were evaluated with contingency tables, where the rain gauges are the "ground truth" values and the MPE are the estimated values. In this way, the accuracy of the rainfall detection in terms of hit rate "H", probability of detection "POD", false-alarm rate "FAR" and discrete bias "DB" can be evaluated.

Table 4.1 shows an example of a two-way contingency table. The variable "a" is the number of times that the rain gauge identifies a rainfall

TABLE 4.1   Two-Way Contingency Table

		Observed Rainfall (Rain gauges)	
		Yes	No
Estimated MPE Rainfall	Yes	a	b
	No	c	c

event and the estimator also correctly identifies a rainfall event at the same time and space. The variable "d" represents the number of times the rain gauge does not observe a rainfall event and the estimator correctly determines that there is no rainfall event. The variable "b" indicates the number of times the rain gauge does not observe a rainfall event but the estimator incorrectly indicates that there is a rainfall event. The variable "c" shows the number of times that the rain gauge detects a rainfall event but the estimator fails to detect the rainfall event [78].

Hit rate (H) is the fraction of the estimating occasions when the categorical estimation correctly determines the occurrence of rainfall event or nonevent. Probability of detection (POD) is the likelihood that the event would be estimated, given that it occurred. The false-alarm rate (FAR) is the proportion of estimated rainfall events that fail to materialize. Bias is the ratio of the number of estimated rainfall events to the number of observed events [142]. The typical scores that measure the accuracy of categorical estimation are:

$$H = \frac{a+d}{n_0} \tag{2}$$

$$POD = \frac{b}{a+b} \tag{3}$$

$$FAR = \frac{b}{a+b} \tag{4}$$

$$DB = \frac{a+b}{a+c} \tag{5}$$

where, $n_o = a + b + c + d$.

The mean field bias (*Bias*) is used to remove systematic error from radar estimates and used to correct the radar quantifications in the hydrologic simulation. The mean field bias is defined as the ratio of the "true" mean areal rain gauge rainfall to the corresponding radar rainfall accumulations [24, 128]. The average of the rain gauge network is evaluated each time step with an arithmetic mean, because the area weights change in time according to malfunctions errors in some gauges. The mean MPE rainfall at each time step is calculated using the area weights as stated above.

The indicators to evaluate the accuracy of MPE rainfall estimations over the HE pixel at different temporal scales are the Bias and root mean square error (*RMSE*).

$$Bias = \frac{\sum_{i=1}^{N_t} G_i}{\sum_{i=1}^{N_t} R_i} \qquad (6)$$

$$RMSE = \left(\frac{1}{N_t} \sum_{i=1}^{N_t} (G_i - R_i)^2 \right)^{\frac{1}{2}} \qquad (7)$$

where, N_t is the number of hours, G_i is the areal mean rain gauge-based rain rate value at time "i", and R_i is the corresponding areal mean radar rain rate value.

For MPE Pixel 1, the associated rain gauges are: C01, C02, C03, C06, C07, C11, L01, L02, L05, L06 and L09, and for MPE Pixel 2 the associated rain gauges are: C04, C05, C08, C09, C10, C12, L03, L04, L07, L08, L10, L11. A mean field bias was calculated at 1 h time resolution. Percentage of rainfall detection by rain gauges and MPE were calculated, and divided into three categories:

- Rainfall not detected by MPE in percent, referred to as "No Radar Detection" or "c".
- Rainfall not detected by rain gauges in percent, referred as "No Rain gauge Detection" or "b".
- Rainfall detected by both sensors in percent, referred as "Coincident" or "a".

The gauges L06 and L08 showed systematic errors in the records and, therefore, were ignored in the calculations. In addition to the statistics computed in the MPE Pixel 1 and MPE Pixel 2, calculations were made using the 4 MPE pixels and the 26 rain gauges for hourly, daily and monthly data accumulations. The PDF was calculated to represent the probability distribution of the daily bias which represents the average total storm correction along one year.

4.5.2 EVALUATION OF FLOW RESPONSE TO RAINFALL INTERPOLATION METHODS

Different interpolation methods can be used to predict areal rainfall between rain gauges or areas where nonareal rainfall information exists. It's important to evaluate how different sources and interpolation methods affect the hydrologic response.

Two interpolation methods are analyzed and compared to produce aerial rainfall from existing rain gauges, which are exponential weighted (EW) and inverse distance weighted (IDW) methods. Additionally, NEXRAD rainfall product level 3 was compared with them. The events analyzed were the Tropical Storm Jean, passing over northern Puerto Rico on November 11–16, 2003.

The interpolations between USGS rain gauges were realized at 200 by 200 m cell resolution and 15 min temporal resolution for each event using the ArcGIS tools. The Hydrologic model (Vflo) with the prepared rainfall information and the MBDB model configuration described in Chapter 4 and aggregated to 200×200 m^2 cell resolution was run with each rainfall product at the same resolution.

Analysis of bias quantification (Eq. 5) between rain gauges and radar were generated for each event and graphical comparisons between scenarios were generated.

4.6 EVALUATION OF PREDICTABILITY DUE TO HYDROLOGIC MODEL PARAMETERS AND INPUT RESOLUTIONS AT TBSW

The previous sections describe which parameters, inputs and initial conditions, up-scaling and interpolation methods can be expected to affect runoff prediction and a hydrologic distributed model in mountainous tropical subwatersheds. With the evolution of instruments to sense the atmosphere (CASA radars, NEXRAD, HE and others), as well as distributed hydrologic models that can predict runoff at even smaller scales, it is necessary to evaluate how the combined effect of model inputs and parameter uncertainties at different scales are spread though the hydrologic model and its impact on reliable operational flood prediction.

The hydrologic evaluation methodology must be objective and unbiased towards a given rainfall input or hydrologic model resolution. Global optimization methods in model calibration seek a unique parameter set that best simulate the observed behavior and if the rainfall resolution or rainfall source is changed, Gourley and Vieux [38] indicated that the model needs to be recalibrated. They proposed a methodology to evaluate the accuracy of the inputs at the hydrologic scale using a hydrologic ensemble. Computing probabilities by examination of the allowable parameter space for each quantitative precipitation estimation algorithm, independently and thus remain unbiased towards a given rainfall source. Model parameter ensembles are created for each rainfall input, the spread and accuracy of the compilation of individual simulations are determined based on comparisons with observed streamflow.

An extension of this methodology will be addressed in this research to include the uncertainties associated with the parameter scale-dependence, in order to determine the accuracy of a given hydrologic model resolution. The combined effect of model parameters, rainfall and model resolution uncertainties are evaluated to produce the predictability limits, computing probabilities by examination of the allowable parameter space for each hydrologic scale and rainfall resolution in combination using ensemble predictions. The TBSW is the ideal scenario to evaluate the predictability limits where a network of rainfall sensors and a flow meter were installed in order to produce rainfall estimates at different scales and then compare the hydrologic prediction to observations for this research.

4.6.1 ESTIMATION OF UNCERTAINTY DUE TO HYDROLOGIC MODEL AT TBSW

Distributed hydrologic model configurations evaluated in this study are applied to represent the real world without any acknowledgment of how they affect the hydrologic prediction and how these uncertainties are propagated in the model at small upland watersheds. It has been shown before that at MBDA indicates input and parameters to be most sensitive in the model, which were used to be tested at the TBSW.

The DEM-derived parameters are well defined for each configuration and are scale-dependent, because they are mainly related to scale issues and aggregation techniques. This type of parameter include: flow accumulation; flow direction; slope; and stream definition indicating implicitly the stream density (as channel cells and overland cells).

The infiltration parameters depend on field measurements of soils and are treated as polygons representations on a map. The soil maps are available for Puerto Rico [USDA, 106–109] and infiltration point measurements are attached to the polygons with the most probable realistic value to represent the area. The polygons are converted to gridded information and, therefore, become scale-depend. The same applies to the roughness map which is, related to the uncertainties associated with the remote sensing techniques, and a probable "realistic roughness value" is used to represent the land use. An up-scaling to the hydrologic model resolution will be addressed to evaluate the effect of parameter uncertainties due to scale.

The effect of slope degradation in the flow quantification was not evaluated. Instead, the aggregation methodology was used to preserve the average slope in the model and decrease the uncertainty and errors due to slope reduction.

The hydrologic evaluation of the resolution models was addressed using parameters ensembles at different resolutions. Every hydrologic parameter was calculated to 50×50 m^2, 100×100 m^2, 200×200 m^2 and 400×400 m^2 resolution from the high-resolution hydrologic model at 10×10 m. The hydrologic evaluation consists of making multiple runs using sets of parameters tested within their distribution's physical bounds and the combinations of inputs for each hydrologic model. Some parameters, such as saturated hydraulic conductivity (Ks), Manning roughness coefficient (n) and initial degree of soil saturation (θ) will be perturbed within their known space, while preserving the spatial variability at a determined scale.

The hydrologically distributed model (Vflo), controls this sampling space by multiplicative factors as illustrated by Moreda and Vieux [126] in the OPPA method that is used to calibrate a distributed model. When no information is known *a priori* about the parameter distributions, uniform distribution is assumed. The scalar factors used to perturb the parameter maps (saturated hydraulic conductivity, Manning roughness coefficient are determined by the following function, which permits computation of probabilities by examination of the allowable parameter space:

$$N_i = \frac{1}{8}(2 + 3i)\big|_{i=0,2,3,4} \tag{8}$$

where, N_i is the adjustment factor [126].

The initial saturation parameter was tested with factor values of 0.25% (dry), 0.4, 0.6, 0.8 and 0.95% (almost fully saturated) covering a sample of the possible parameter space. Vieux and Vieux [132, 133] tested a long-term distributed model at Loiza, Puerto Rico and found initial saturation factors around 0.75 in the uncalibrated model and 0.9 in the calibrated model. Additionally the initial soil saturation did not fall below 0.25 in the run time.

Each initial condition (rainfall event and one hydrologic setting resolution) and parameter perturbation was run in the hydrologic model (Vflo) producing a deterministic prediction called "ensemble member", which are treated collectively and are samples of the PDF, representing the true initial state distribution. The three-parameter perturbation in combination with one determined hydrologic and rainfall resolution event will produce a hydrologic ensemble. Each ensemble required 125 *Vflo* runs or ensemble members obtaining a simulation sample space for each hydrologic resolution model and rainfalls are stored in a separate folder.

Results of each simulation were compared to the observed streamflow at the TBSW outlet. Three variables are important to evaluate in a flash flood forecasting, providing information of the flood magnitude (peak to flood), spread (volume normalized by the area) and lead time (time to peak) for the emergency management agencies. Box plots of each ensemble permit visualization of the spread of the solution due to parameters perturbations at each rainfall and model scale.

The estimation of uncertainty due to hydrologic model up-scaling was performed regrouping the ensembles mentioned. The ensembles here are formed by the perturbations of the parameters and rainfall resolutions. Then, a hydrologic model resolution is evaluated according its size and is not dependent on rainfall resolution, because, it is tested with all rainfall resolutions. An important tool for the modeler is to understand the implications of using one specific hydrologic model resolution to estimate the flow discharge reliably.

Different objective functions exist, such as the least square error or maximum likelihood, to evaluate the variables in a verification step. The least square error is computed for each streamflow prediction giving a better understanding of the shape of the hydrograph.

The forecast or prediction verification method of an ensemble is the process of assessing the quality of the prediction with the corresponding observation. The quantitative statistics provide a simple way to evaluate the quality of an ensemble. To average the members of the ensemble to obtain a single prediction, provide a prediction that is more accurate than the single prediction initialized with the best estimate of the initial state of the hydrologic parameters. The mean ensemble is an overall indicator of the ensemble's behavior and is considered to be the best estimate [99].

The spread skill relationship for a collection of ensemble forecasts often is characterized by the correlation between the variance o the square of the standard deviation of the ensembles members around their ensemble mean. The accuracy is often characterized using the mean squared error.

The mean Time, Peak and Volume of each ensemble is computed and compared with observations. Additionally, the following statistics were used: *Bias*, Mean Absolute Error (*MAE*) and Root Mean Square Error (*RMSE*). These definitions are formulated below:

$$Bias = E\left[y_k\right]/O \tag{9}$$

$$MAE = \frac{1}{n}\Sigma_{K=1}^{n}\left|y_k - O\right| \tag{10}$$

$$RMSE = \sqrt{\frac{1}{n}\Sigma_{K=1}^{n}\left(y_k - O\right)^2} \tag{11}$$

where, y represents the prediction from the k-th simulation for Time, Peak and Volume, and O is the observation.

The *Bias* measures the correspondence between the average forecast and the average observed value of the predictands. The *MAE* is the arithmetic average of the absolute values of the differences between the members of each pair. The *MAE* and *RMSE* values near to zero are desirable while *Bias* near to one are expected.

Another diagnostic variable for representing runoff generation is the runoff coefficient that is equal to observed discharge volume divided by the basin-average rainfall event. These spread skill correlations have been found to be fairly modest, accounting for 25% or less of the accuracy variations [5, 42, 44]. Alternative approaches to the spread skill problem using probability distributions for forecast skill, conditional on ensemble spread were analyzed by Moore and Kleeman [69]. The conditional PDF are a statistical tool more robust than a simple ensemble mean to compare to an observation. PDF's were calculated for Time to Peak, Volume and Peak flow using the 625 ensemble members for the combination of hydrologic resolution model and rainfall event. The most widely used and important continuous probability distribution is the Gaussian or normal distribution described as:

$$p_x(x) = \frac{1}{\sqrt{2\pi\sigma^2}} e^{\frac{-1}{2}\left(\frac{x-\mu}{\sigma}\right)^2} \tag{12}$$

where, μ and σ^2 the mean and the variance of X, respectively.

Thus, the normal distribution is a two-parameter distribution which is bell-shaped, continuous, and symmetrical about the mean.

With the PDF, measures of the central tendency, prediction spread, limits and skill can be estimated. The central tendency is represented by the 50% simulation limit, or median, corresponding to 0.5 on the cumulative distribution function (CDF). The spread of the forecast represents the forecast uncertainty due to uncertain initial conditions, rainfall inputs, slopes and scale dependent parameters, etc.; by determining the distance between the 5% and 95% confident limit simulation bounds.

The ensemble skill is assessed using the ranked probability score, RPS [29, 71] which is capable of penalizing forecasts increasingly as more probability is assigned to event categories further removed from the actual outcome and the ensemble are encouraged to report their "true beliefs" [142]. Brier scores and reliability diagrams are used to evaluate each of the derived binary forecasting situations, but the RPS is an option for verification forecasts for multi category ordinal predictands.

The ranked probability score is the sum of squared differences between the components of the cumulative forecast and observation vectors as:

$$RPS = \sum_{m=1}^{J} \left(Y_m - O_m\right)^2 \tag{13}$$

$$RPS = \sum_{m=1}^{J} \left[\left(\sum_{j=1}^{m} y_j\right) - \left(\sum_{j=1}^{m} O_j\right)\right]^2 \text{, and} \tag{14}$$

$$Y_m + O_m + = 1 \text{ always}$$

where, Y_m and O_m are the cumulative forecast and observation, respectively, y_j is the cumulative probability assigned to the category or vector component, o_j is the cumulative probability of the observation in the ith category or vector component and J is the number of categories and therefore also the number of probabilities included in each forecast. The sum of Y_m and O_m are always both equal to one by definition.

The PDFs statistics and RPS generated for each grid size will contain the predictability limits for small watersheds and will be useful information that can help the modeler to decide which grid size resolution is appropriate for larger watersheds where it is important to quantify flash flooding at upstream and ungauged sites.

The Figure 4.3 summarizes the evaluation of uncertainty propagation though flow prediction. The flow chart used a combination of hydrologic parameter perturbations within the physical bounds, rainfall input and model resolution or structure set up.

Knowing the uncertainty at the small scale and associated with the resolution selection, it will produce more realistic parameter estimations and flood quantification for the larger scale model. In other words, if the small scale, high-resolution model, is characterized by a degree of uncertainty, then the goal of the modeler is to up-scale the resolutions, while maintaining a similar degree of uncertainty. In this way, the modeler hopes to maintain accuracy at the subwatershed scale.

4.6.2   ESTIMATION OF UNCERTAINTY DUE TO RAINFALL UP-SCALING AND TEMPORAL VARIATIONS

The same methodology, described in section 4.6.1 in this chapter, was used to calculate the uncertainty due to rainfall up-scaling and temporal

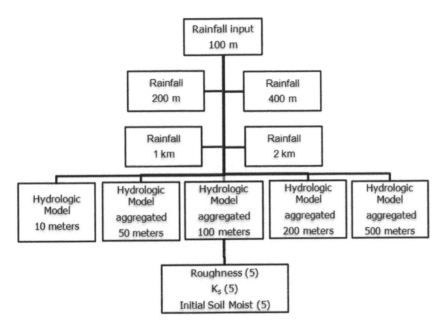

FIGURE 4.3   Flow chart of the ensemble for predictability limits.

variations. The amounts of rainfall measured by the rain gauge network within the TBSW are assumed to represent the "true" rainfall. The rain gauges are the most reliable method to sense precipitation and are widely used to correct other sensors methods (e.g., radar, satellite and laser sensors) and remove sensor bias.

By interpolating to various resolutions, it is possible to measure the importance of spatial rainfall variation in hydrologic prediction while the average rainfall falling on the watershed is maintained, taking into account that the average distance between the rain gauges is approximately 218 m with a standard deviation of 100 m.

Precipitation total variations between rain gauges were calculated and presented for each event, demonstrating the high rainfall variability at small scales due to orographic effects in mountainous subwatersheds. The rainfall events were interpolated to the following resolutions: 100 m, 200 m, 400 m, 1000 m, and 2000 m to compare them in a probabilistic and deterministic sense. The interpolation method used was the inverse distance method. Each ensemble had 625 runs or ensemble members. These were the combination of: parameter perturbations (125 runs), model structures (5 different model resolutions), and one rainfall event (Figure 4.3). Observed and simulated values were compared by using objective functions. The compared variables were time to peak, peak flows and volume.

In addition, PDFs were computed using the Gaussian kernel density estimation technique and computation of nonparametric statistics provided information for the 0.05, 0.5 and 0.95 quartiles, given the central tendency and spread of the ensemble. The PDFs are treated as conditional probabilities and not as the true probability distribution. RPS's were calculated to compare the skill of each rainfall input. Rainfall events were tested though the year using different antecedent soil moisture conditions and temporal patterns. The dates tested were: October 22, 2007; May 2, 2008; June 5, 2008; August 28, 2008 and September 3, 2008. Performing the statistics previously described for each rainfall configuration ensemble, it was possible to evaluate the reliability of one rainfall resolution and compare them event-by-event and assess if there exists variations between events.

CHAPTER 5

FLOOD PREDICTION LIMITATIONS IN SMALL WATERSHEDS: SENSITIVITY ANALYSIS[1, 2]

ALEJANDRA M. ROJAS-GONZÁLEZ

CONTENTS

5.1 INTRODUCTION

This chapter includes results for the sensitivity analysis performed in the MBDB (see, Section 5.2) for different hydrologic parameters and rainfall input. Spider plots for percentage changes in peak flow; and runoff depth versus scalar factors (0.5, 1, 2.5 and 2) were plotted. Additionally, relative sensitivity coefficient analysis was addressed for ± 50% of parameter and input change (or 0.5 and 1.5 multiplicative factors). The most sensitivity

---

[1] This chapter is an edited version from, *"Alejandra María Rojas González, 2012. Flood prediction limitations in small watersheds with mountainous terrain and high rainfall variability. Unpublished PhD Thesis for Department of Civil Engineering and Surveying, University of Puerto Rico – Mayagüez Campus"*.

[2] Numbers in brackets refer to the references at the end of this book.

parameters found were used in the up-scaling experiment to be perturbed in the TBSW. Section 5.3 describes the methods to fill the gaps between rain gauges and radar data in the MBDB.

5.2   PARAMETERS AND INPUT SENSITIVITY: SALIENT FINDINGS

To identify the parameters for which the MBDB model is most sensitive for the mountainous condition considered, a sensitivity analysis was conducted. Uncertainties associated with the model parameters and inputs can be quantified by evaluating the hydrologic response given a range of parameter and input perturbations at 0.5, 1, 1.5 and 2 multiplicative factors or scalars. Within the study area, 3 USGS flow stations were identified, Río Grande de Añasco near San Sebastian, Río Guanajibo near Hormigueros and Río Rosario near Hormigueros. The parameters within the drainage area upstream of the USGS flow stations were perturbed by the multiplicative factors conserving the spatial distribution. Sets of parameter used in the hydrologic model were shown in Tables 5.4 and 5.6 as well as very shallow soil depth (20 cm); and initial saturation fraction of 0.5 was selected as a preliminary hydrologic model configuration at 200 m resolution.

The rainfall was created using additional USGS stations upon availability for each event. The point rainfall estimates at 15 min were interpolated at 200 m resolution using the exponential weighted interpolation. For hurricane Georges (September 21 to 23, 1998) only three USGS stations mentioned above were working. For November 11 to 16, 2003 event, eight USGS station were interpolated and for September 14 to 17, 2004 seven stations. Figure 5.1 shows the storm total maps for the interpolations performed for each rainfall event at 200 m resolution using the stations available; the dots within each figure are the station locations with data each 15 min. The maximum rainfall accumulation during each event was 566.5 mm for September (Figure 5.1A), 291.6 mm for November, 2003 (Figure 5.1B), and 156.2 mm for September, 2004 (Figure 5.1C).

Spider plots were drawn for the parameters and rainfall perturbed additionally, relative sensitivity coefficients (Sr, Eq. (1) in Chapter 3) were

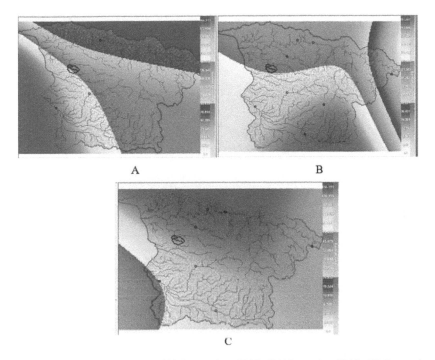

FIGURE 5.1   Total storm maps, (A) September, 1998; (B) November 2003; (C) September 2004.

calculated with changes of ±50% using the hydrologic distributed model for three events mentioned and three outlet points; considering the behavior of two output variables (runoff depth and peak discharge).

Spider plots are used to evaluate the model response to the entire range of the parameter and determine if there is a portion of the parameter range that yields unrealistic results. Figure 5.2 presents the spider plots for peak flow as percent change in the model output variable versus change in rainfall value by a multiplicative given factor. Variations in the hydrologic response are linear; doubling the rainfall input increase the peak flow from 131.7% to 203.2% for Río Guanajibo near Hormigueros depending on the rainfall event. In the case of Río Grande de Añasco near San Sebastian the range is between 135.3% and 168.5% and for Río Rosario near Hormigueros is between 127.7% and 145.3%.

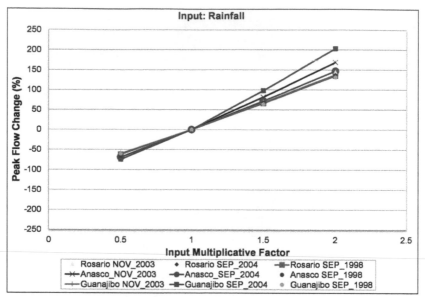

FIGURE 5.2 Spider plot for percentage change in peak flow due to rainfall multiplicative factors at 3 USGS station outputs.

Figure 5.3 presents the spider plot for runoff depth where the linearity between rainfall perturbations and hydrologic response was not conserved. For example, doubling rainfall generates a runoff depth change between 111.5% and 145% for Guanajibo and 131.4% and 135.0% for Añasco; and between 112.4% and 120.6% for Rosario. These results indicate that the infiltration is decreased with increasing the rainfall intensity providing the volume to the runoff that could not be infiltrated. Decreasing the rainfall intensity by 0.5 multiplicative factors, favors infiltration and decreases the runoff depth with percent changes between 25.5% and 64.8%. Lower percentages are presented for September 2004 (25.5% – 31.8%), which has a rainfall pattern different from the others (Figure 5.1C). This event is characterized by high rainfall intensity (red color) in the upland and lower in the flood plains. Minor percent variations occur with the peak flow for Añasco and Rosario discharge points (61.9% to 69.1%) compared with Guanajibo (50% to 74%).

Increasing channel roughness decreased the peak flow (Figure 5.4C), while increasing initial soil saturation increased the peak flow (Figure 5.4A),

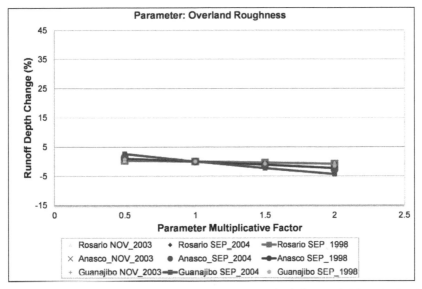

FIGURE 5.3   Spider plots for percentage change in runoff depth due to rainfall multiplicative factors at 3 USGS station outputs.

especially in Río Guanajibo near Hormigueros outlet point, for September, 2004. Low variations were founded in peak flow with variations of soil depth and hydraulic conductivity for all events (Figures 5.4B, 5.4E, 5.4F).

Additionally, spider plots graphs for runoff depth changes were drawn and presented in Figure 5.5 for each parameter under evaluation. As for peak flow, percent changes were graphed for different events and outlet points. The parameter that produced the greatest percentage change in runoff depth was the initial soil saturation (Figure 5.5A), for Añasco near San Sebastian outlet point for November 2003 and September 1998 and Guanajibo near Hormigueros for September, 2004. Generating a change between 30% and 40% in runoff depth due to doubling in the initial soil saturation, where the baseline was 0.5 and doubling produced a value of 1 (i.e., saturated conditions). Low variations were found with changes of the other parameters (Figures 5.5B–5.5F). The magnitude of change varied with the event indicating that the rainfall spatial distribution and intensity are important aspects for quantification of initial parameters.

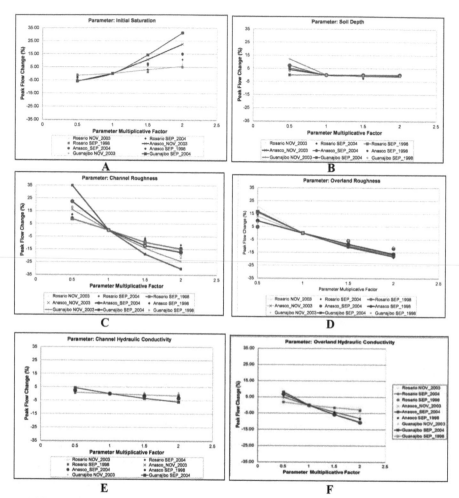

FIGURE 5.4   Spider plots for changes in peak flow due to parameters multiplicative factors evaluated at USGS stations and 3 events. Parameters: A) Initial Saturation, B) Soil Depth, C) Channel Roughness, D) Overland Roughness, E) Channel hydraulic conductivity, F) Overland hydraulic conductivity.

Relative sensitivity coefficients were calculated for parameters and rainfall input using each event and outlet point. Results are presented in Table 5.1 for the peak flows and Table 5.2 for runoff depth as well as averages and standard deviations.

Results given below indicate that variations for both output variables (peak flow and runoff depth) are most sensitive to the rainfall input with a

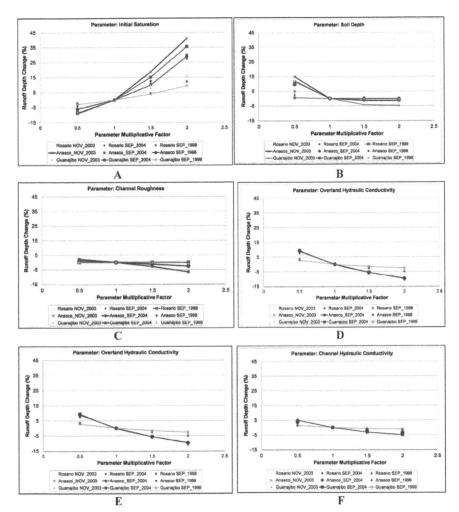

FIGURE 5.5 Spider plots for changes in runoff depth due to parameter multiplicative factors evaluated at USGS stations and 3 events. Parameters: A) Initial Saturation, B) Soil Depth, C) Channel Roughness, D) Overland Roughness, E) Channel hydraulic conductivity, F) Overland hydraulic conductivity.

Sr of 69.1 and 56.5, respectively. Runoff depth was affected by initial saturation, increases in this parameter increased the runoff and a Sr value of 8.2 was obtained. Followed by overland hydraulic conductivity with a Sr of −5.5, increase in this parameter decreased the runoff depth; and increasing soil depth produced a decrease in peak flows (Sr of −4.4). Low variations

TABLE 5.1 Relative Sensitivity Analysis for Peak Flow Evaluating Three Events and Three USGS Station Outlet Points for Peak Flow

	Rosario				Añasco				Guanajibo					
	Nov 03	Sep 04	Sep 98	Mean	Nov 03	Sep 04	Sep 98	Mean	Nov 03	Sep 04	Sep 98	Mean	Average Sr	STD
Rainfall	66.85	66.86	63.87	65.9	75.3	70.4	66.7	70.8	63.8	86.1	62.1	70.7	69.1	7.5
Rough Ch	−5.07	−8.13	−9.14	−7.4	−14.3	−17.3	−9.8	−13.8	−15.0	−26.9	−15.2	−19.0	−13.4	6.4
Slope	10.76	10.20	11.85	10.9	13.1	11.9	10.5	11.8	12.7	20.6	11.9	15.1	12.6	3.1
Rough over	−15.47	−12.55	−13.07	−13.7	−11.1	−5.3	−9.0	−8.5	−8.1	−13.7	−6.8	−9.5	−10.6	3.4
IS	4.42	2.75	1.18	2.8	8.3	6.4	5.1	6.6	5.9	10.1	2.3	6.1	5.2	2.9
K_s Over	−3.52	−4.55	−1.89	−3.3	−6.8	−6.3	−5.3	−6.1	−3.4	−7.0	−1.7	−4.0	−4.5	1.9
Soil Depth	−0.10	0.00	−0.03	0.0	−2.5	−3.6	−3.1	−3.0	−6.5	−2.6	−1.5	−3.6	−2.2	2.1
K_s Chan	−0.97	−1.07	−0.44	−0.8	−2.5	−2.1	−1.4	−2.0	−0.9	−3.9	−0.8	−1.9	−1.5	1.1

IS = initial saturation, K_s Over = overland hydraulic conductivity; K_s Chan = channel hydraulic conductivity; Rough Ch = channel roughness; Rough over = overland roughness; S Depth = soil depth.

TABLE 5.2 Relative Sensitivity Analysis for 3 Events and 3 USGS Station Outlet Points for Runoff Depth

	Rosario (Month, year)				Añasco (Month, year)				Guanajibo (Month, year)				Average	STD
	Nov 03	Sep 04	Sep 98	Mean	Nov 03	Sep 04	Sep 98	Mean	Nov 03	Sep 04	Sep 98	Mean		
Rainfall	60.00	49.22	55.82	55.0	67.5	49.2	64.1	60.3	58.6	48.7	55.5	54.3	56.5	6.74
IS	8.75	7.55	3.18	6.5	13.8	9.9	8.0	10.6	7.0	11.9	3.5	7.5	8.2	3.50
K_s Over	-5.57	-6.42	-3.04	-3.9	-8.1	-7.4	-6.7	-7.4	-3.1	-7.2	-2.2	-4.2	-5.5	2.20
S Depth	-6.04	-0.86	-0.28	-2.4	-8.1	-5.3	-2.6	-5.3	-8.2	-5.8	-2.3	-5.4	-4.4	2.98
K_s Chan	-2.28	-2.57	1.24	-1.6	-3.4	-3.0	-2.7	-3.0	-1.2	-4.1	-1.2	-2.2	-2.4	1.30
Slope	0.39	0.88	0.28	0.5	1.4	2.2	1.0	1.5	0.9	3.9	0.9	1.9	1.3	1.12
Rough over	-0.37	-1.06	-0.35	-0.5	-0.7	-1.5	-1.1	-1.1	-0.5	-2.5	-0.8	-1.3	-1.0	0.67
Rough Ch	-0.09	-0.15	-0.05	-0.1	-0.6	-1.2	-0.6	-0.8	-0.6	-2.5	-0.4	-1.2	-0.7	0.77

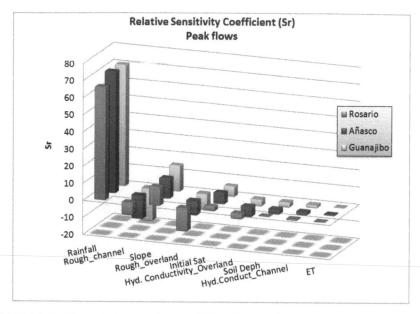

FIGURE 5.6   Mean relative sensitivity coefficients for peak flows at three USGS outlet points.

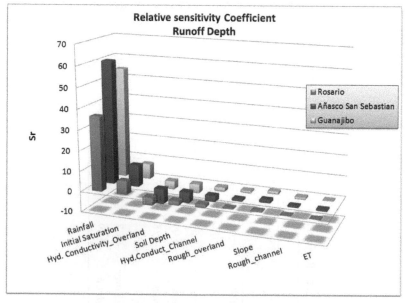

FIGURE 5.7   Mean relative sensitivity coefficient for runoff depth at three USGS outlet points.

were observed when soil depth was doubled, indicating that soil depths greater than 40 cm will produce little runoff depth changes (Figure 5.4B).

The peak discharge was affected by roughness with a Sr of -13.4 for channel cells and Sr of -10.6 for overland cells; increases in roughness parameter decreased the peak flows and retarded the time to peak. The slope-distributed map produced a Sr of 12.6, increasing this parameter increased peak flow. The initial soil saturation parameter produced a Sr of 5.2 and is placed in the fifth place. Average relative sensitivities coefficients (Tables 5.1 and 5.2) were plotted in Figures 5.6 and 5.7 with observed variations in terms of basin outlet points or events.

5.3 SENSITIVITY DUE TO QUANTITATIVE PRECIPITATION ESTIMATION WITHIN GAP AREAS

The *Vflo* model has the capability to support distributed rainfall and rain gauge data in real time, ideal for a flood alarm system. However, rainfall itself is the principal source of uncertainty in the model as observed in the previous section. The number of rain gauges in a basin are frequently sparse and therefore do not capture the spatial variability.

Two interpolation methods, exponential weighted (EW, Figure 5.8A), and inverse distance weighted (IDW, Figure 5.8B), were compared with radar rainfall from NEXRAD level 3 as seen in Figure 5.8C, for the November 11–16, 2003 period. The average total storm rainfall calculated at an outlet point is different between interpolation methods and radar source. For example for the USGS station Río Grande de Añasco near San Sebastian the precipitation average depth is 122.8 mm for IDW, 114.8 mm and for EW and 77.8 mm for radar. In the USGS station at Río Guanajibo near Hormigueros, the total storm was 230.6 mm with IDW, 237.1 mm and for EW and 199.8 mm for radar.

It should be noted that the radar is partially dependent on the rain gauge data and number of stations. Furthermore, when we use radar, it is necessary to remove systematic error by applying a calculated correction factor or bias [129] for the event, which is the relationship between rain gauges and the radar data. For November 2003 event, the bias calculated for the whole area was 1.3 (Eq. (5) in Chapter 4). Figure 5.9 displays the scatter plot of radar and rain gauges and the adjusted line.

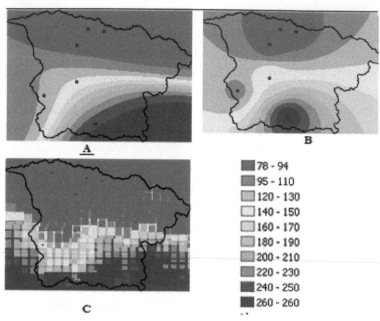

FIGURE 5.8 Total Storm Rainfall Maps at Mayagüez Bay Drainage Basin for November 11–16, 2003 using Interpolation Methods: (A) Exponential Weighted; (B) Inverse Distance Weighted; and Radar data (C).

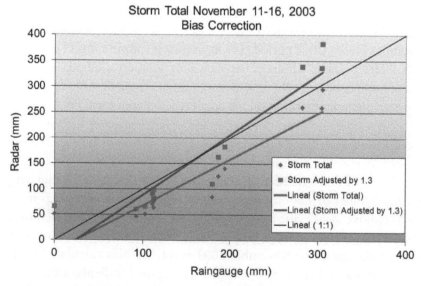

FIGURE 5.9 Radar Bias correction for storm total, November 11–16, 2003.

Variations between methods to fill the gaps between rain gauges pro-duce different responses in flow prediction. For example for the MBDB model we performed hydrologic simulations using the EW and IDW inter-polation methods at 200 m resolution and NEXRAD radar level 3 at 2 km spatial resolution with a nominal resolution of 500 m. The results were compared at Río Grande de Añasco near San Sebastian and Guanajibo near Hormigueros stations generating differences in peak flow runoff depth and average total rainfall (Table 5.3).

The EW method produced greater peaks (2.4%) and runoff depth (2.5%) at Guanajibo outlet point, with a decrease in rainfall total storm (2.9%) than IDW. The reverse effect was observed at Río Grande de Añasco where decreasing the rainfall total rainfall (–6.5%) generated proportional decrease in peak flow (–7.1%) and runoff depth (–6.8%). The radar rainfall quantification is –12.9% and –36.7% lower than IDW for Guanajibo and Añasco, respectively, however the reduction in peak flow was not in the same proportion indicating that the rainfall intensity was maintained.

TABLE 5.3 Comparison of Hydrologic Results and Rainfall Interpolation Methods and Radar

	Río Guanajibo near Hormigueros					
	Peak Flow		Runoff depth		Rainfall	
	(CMS)	Percent change	(mm)	Percent change	(mm)	Percent change
IDW	394.1	Reference	145.9	Reference	230.6	Reference
EW	403.4	2.4	149.6	2.5	237.1	2.9
Radar	376.6	–4.4	128.5	–11.9	200.9	–12.9
	Río Grande de Añasco near San Sebastián					
IDW	668.4	Reference	117.6	Reference	122.8	Reference
EW	620.9	–7.1	109.6	–6.8	114.8	–6.5
Radar	642.8	–3.8	72.4	–38.5	77.8	–36.7

CHAPTER 6

FLOOD PREDICTION LIMITATIONS IN SMALL WATERSHEDS: BIAS ESTIMATION IN RADAR PRECIPITATION PRODUCT[1, 2]

ALEJANDRA M. ROJAS-GONZÁLEZ

CONTENTS

6.1 INTRODUCTION

In this chapter, an analysis of the rainfall spatial variability in a small area with a high-density rain gauge network is described. Radar rainfall estimations were compared and evaluated with the rain gauge data. Statistical measurements of discrete and continuous validation scores were calculated for the radar estimates at hourly and daily time step. PDFs were calculated for the Bias with the purpose of knowing the rainfall uncertainty over a small area.

---

[1] This chapter is an edited version from, *"Alejandra María Rojas González, 2012. Flood prediction limitations in small watersheds with mountainous terrain and high rainfall variability. Unpublished PhD Thesis for Department of Civil Engineering and Surveying, University of Puerto Rico – Mayagüez Campus"*.

[2] Numbers in brackets refer to the references at the end of this book.

6.2   BIAS ESTIMATION IN RADAR PRECIPITATION PRODUCT

To compare the *Multisensor Precipitation Estimates* (MPE) with the rain gauge network rainfall accumulation time series, it is necessary to convert the MPE HAP grid projection to a State Plane raster product, which will be used in the hydrological model. Due to changes in coordinates and raster conversions, the original pixels (HAP projection) oriented with a certain angle, were reoriented horizontally (raster).

Figure 6.1 displays the change in the orientation, including the MPE pixels (left) and Hourly Rainfall Product (N1P) from NEXRAD level 3 (right). The left image shows four square black boxes corresponding to the MPE raster-projected pixels, the colored pixels are the original raster with HAP coordinates at 4×4 km² spatial resolution, and the red box corresponds to the Hydro-Estimator pixel at the same resolution as the MPE product.

The annual 2007 rainfall accumulations for the 4 MPE pixels were 1546.2, 2212.1, 1949.8 and 2088.6 mm, with an annual standard deviation of 289.3 mm between them. Figure 6.2 shows the temporal variations in the cumulative rainfall during the year for each MPE Pixel. Large differences are found between Pixel 1 and Pixel 2.

6.2.1   MONTHLY CUMULATIVE RAINFALL

To show how variable the rainfall distribution within a specific pixel can be, authors took the MPE Pixel numbers 1 and 2 and determined the

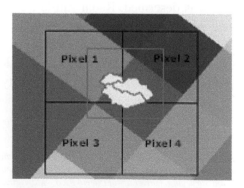 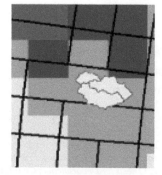

FIGURE 6.1   HE pixel (red box) and MPE pixels (black and colored boxes) (left) and Hourly Rainfall Product (N1P) from NEXRAD level 3 (right) orientated in shapefile and raster formats.

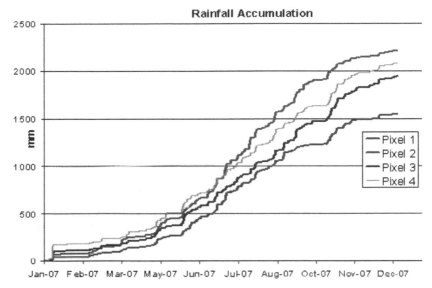

FIGURE 6.2 Rainfall accumulations over the time for the MPE pixels.

rain gauges associated with each pixel. A plot of the monthly cumulative rainfall for MPE Pixel 1 and rain gauges are displayed in Figure 6.3. The cumulative rainfall for the months of April and May are not representative of those months because we had missing rain gauge data for 11 days for April and 9 days for May, therefore, the computations were made with only the available data for these months. For the case of July, Figure 6.3 shows that only the C06 station reported an amount of rainfall (206.9 mm) that was similar to the MPE Pixel 1 rainfall (259.15 mm), and for almost all months, note that the MPE Pixel 1 underestimated the rainfall value with respect to rain gauges, except for the months of January, June and July.

6.2.2 AVERAGE RAIN GAUGE NETWORK RAINFALL

Figure 6.4 displays the average rain gauge network rainfall in MPE Pixel 1 versus the standard deviation for 1-hour time step for 2007. The slope between standard deviation and mean rainfall is equivalent to the coefficient of variation (CV), and is a measure of the dispersion of the probability distribution. From the regression analysis, a R^2 of 0.6627 and a CV of

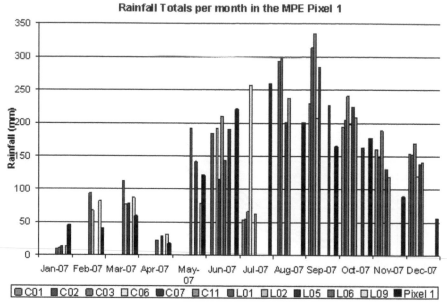

FIGURE 6.3   Monthly Total Rainfall calculation for the rain gauge stations belonging to MPE Pixel 1, for 2007.

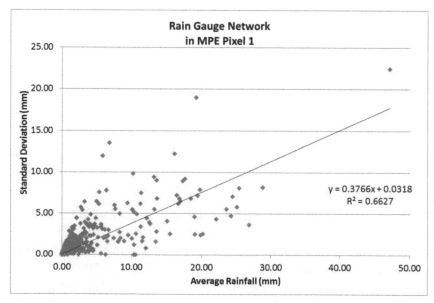

FIGURE 6.4   Hourly average and standard deviation rainfall for the rain gauge network corresponding to MPE pixel 1 for 2007.

0.3766 were obtained, indicating high rainfall variability in the MPE pixel 1, which cover an area of 4.5 km^2.

The rain gauge network covering an area of 16 km^2 shows that the relationship between mean rainfall and standard deviation has the trend of an increase in rainfall depth will produce an increase in standard deviation. The linear regression indicates a R^2 of 0.78 and a slope of 0.45 (Figure 6.5). An increase in CV exists between Figures 6.5 and 6.6, related to an expansion of the rain gauge area from 4.5 to 16 km^2 indicating an increase in dispersion of the data. Therefore, the coefficient of determination increases, indicating that the standard deviation of a sample of mean rainfall can be obtained with more accuracy than in small areas.

Mean rain gauge network data and mean weighted MPE rainfall were graphed at the hourly time step and a linear regression equation was calculated (Figure 6.6) obtaining a slope line of 0.848 and a R^2 of 0.43. The slope represents the Bias between the rainfall from the gauge network and the MPE radar product, and this value can be applied to the hourly MPE measurements as a correction. The MPE in general is overestimating

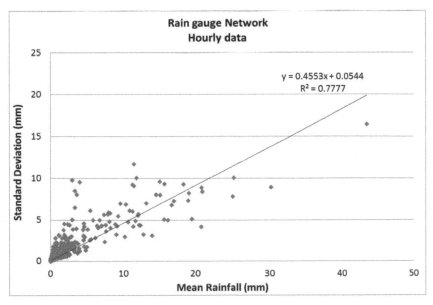

FIGURE 6.5 Hourly average and standard deviation rainfall for rain gauge network for 2007.

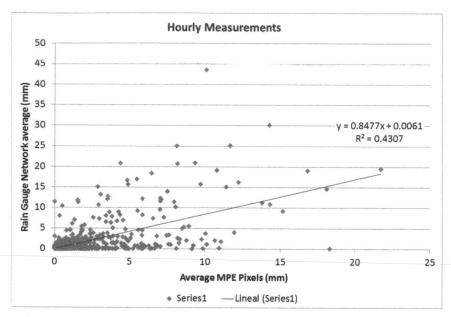

FIGURE 6.6 Average rain gauge rainfall vs. MPE radar rainfall within HE pixel at hourly time step.

precipitation with a coefficient of determination of 0.4307. The MPE exhibits problems of detection at low rainfall measurements principally (Figure 6.6).

6.2.3 CONTINGENCY TABLES AND SCORES

The contingency tables and scores (Tables 6.1 and 6.2, respectively) were calculated to evaluate the Pixel 1, Pixel 2 and total 4 MPE pixels for hourly time step and daily rainfall accumulations for the four MPE pixels within the HE pixel. The number of estimated rainfall events was overestimated according to the discrete bias (DB) in the MPE pixel 1 (1.24) comparing with the Pixel 2 and the 4 MPE pixels, which have a value close to 1. For daily data the DB is underestimated by a factor of 0.956.

The hit rate (H) indicates the occasions when the categorical estimation correctly determined the occurrence of rainfall event or nonevent and was around 0.82 and 0.89; nonsignificant differences were found between hourly and daily accumulations at the 4 pixels.

TABLE 6.1   Contingency Tables for the MPE Pixels

Hourly Data MPE Pixel 1		Observed Rainfall (Rain gauges)	
		Yes	No
Estimated MPE Rainfall	Yes	638	653
	No	400	6581
Hourly Data MPE Pixel 2		Observed Rainfall (Rain gauges)	
		Yes	No
Estimated MPE Rainfall	Yes	630	464
	No	449	6729
Hourly Data 4 MPE Pixels		Observed Rainfall (Rain gauges)	
		Yes	No
Estimated MPE Rainfall	Yes	915	756
	No	693	5910
Daily Data 4 MPE Pixel		Observed Rainfall (Rain gauges)	
		Yes	No
Estimated MPE Rainfall	Yes	225	33
	No	45	341

TABLE 6.2   Discrete Validation Scores for the MPE Pixels and Time Scales

	Hourly Data			Daily Data
	MPE Pixel 1	**MPE Pixel 2**	**4 MPE pixels**	**4 MPE pixels**
POD	0.62	0.58	0.57	0.833
FAR	0.51	0.42	0.45	0.128
DB	1.24	1.01	1.04	0.956
H	0.87	0.89	0.82	0.879

Moreover, the probability of detection (POD) is the likelihood that the event would be estimated by the radar, increasing with the time step, with 0.833 for the daily data. Daily estimates eliminate the influence of light rainfalls that the radar cannot detect. For the hourly time step, the Pixel 1 POD was higher than the POD for Pixel 2 and the average of 4 MPE pixels.

6.2.4 FALSE ALARM RATES OR PORTION OF ESTIMATED RAINFALL EVENTS

False alarm rates or portion of estimated rainfall events that fail to materialize are similar in Pixels 1, 2 (0.50 and 0.42, respectively) and the four pixels average (0.45). For the daily time step there was a considerable reduction in the FAR (0.128). Figures 6.7 and 6.8 show the distribution of false alarms and the probability of no detection by the radar during 2007. Events in which the radar did not detect rainfall and the rain gauges did measure rainfall (c) were assigned a value of 1 in the graph. Events in which the radar did detected rainfall and the gauges did not measure rainfall (b) were assigned a value of 2. Differences in time when false alarms and probability of no detection quantities occurred can be observed in the graphs, and detailed statistics are presented in Tables 6.2 and 6.3.

6.2.5 MEAN FIELD BIAS (BIAS)

A mean field bias (Bias) was calculated for the MPE Pixel 1, 2 and overall 4 pixels, as the ratio of the average of the rain gauge rainfall and the mean

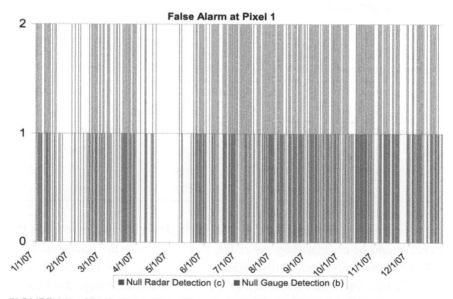

FIGURE 6.7   Hourly False Alarm Time Series for the MPE Pixel 1 for 2007.

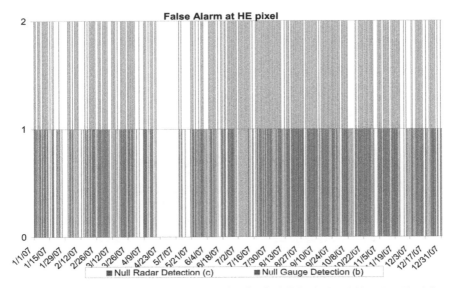

FIGURE 6.8 Hourly False Alarm Time Series for the MPE Pixels within a HE Pixel for June to December 2007.

TABLE 6.3 Continuous Validation Scores for the MPE Pixels and Time Scales

	Mean Hourly				Daily Data
	MPE Pixel 1	MPE Pixel 2	4 MPE pixels	4 MPE pixels Rain ≥ 0.3 mm	4 MPE pixels
RMSE	–	–	0.012	–	0.368
Bias	3.85	1.58	2.77	1.55	1.23
STD *Bias*	4.21	2.73	8.18	2.14	1.65

rainfall sensed for the MPE pixels using the area weights for each time step (hourly, daily, monthly and annually accumulations). Hourly mean field bias time series during the 2007 are displayed in Figure 6.9 for the MPE Pixel 1 only and Figure 6.10 for the mean four MPE pixels within the HE pixel.

Large biases were found at the hourly time step and are associated with small radar rainfall and rain gauge detections (Figure 6.9). The possible effect is that the radar minimum precipitation depth capable of being detected is 0.01 inches or 0.254 mm; while our rain gauge network has

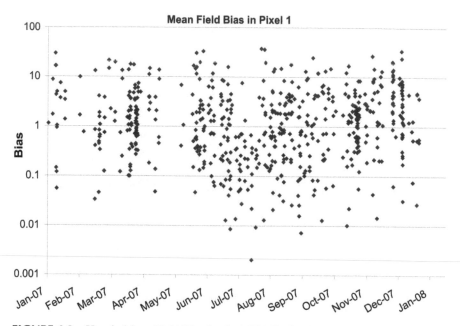

FIGURE 6.9   Hourly Mean Field Bias for the MPE Pixel 1 during 2007.

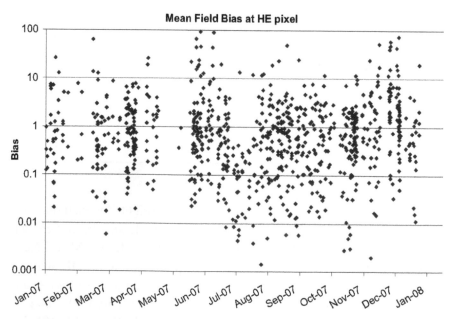

FIGURE 6.10   Hourly Mean Field Bias for the four MPE Pixels during 2007 within a HE Pixel.

a rainfall depth resolution of 0.1 mm. In addition, the NEXRAD in Puerto Rico is located about 100 km from the study area in Cayey at a site elevation of 850 meters amsl. Due to the earth curvature, the beam has an elevation of 600 m above the study site at Mayagüez, affecting the cloud's measurement in the lower troposphere.

To neutralize the noise effect of small rainfall quantifications in the hourly bias computation, rainfall depths less than 0.3 mm were eliminated. A considerable hourly bias reduction was observed in time (Figure 6.11) and in the average and standard deviation computation across the year as well as monthly (Tables 6.3 and 6.4).

The continuous validation scores for MPE rainfall validation (Table 6.3) show a root mean square error is greater (0.368 mm) in daily accumulations than in hourly (0.012 mm). The mean field bias average for 2007 in Pixel 1 is 3.85 with a standard deviation average of 4.21. The four MPE pixels present a lower Bias (2.77) but a large standard deviation (8.18). The annual average Bias is improved after eliminating rainfall depths less than 0.3 mm, diminishing to 1.55 and a standard deviation of 2.14 for the four MPE pixels with rainfall greater than 0.3 mm.

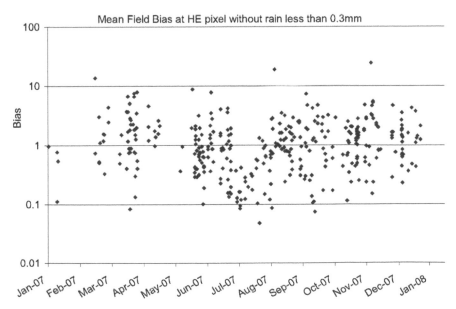

FIGURE 6.11 Hourly Mean Field Bias for the overall MPE Pixels within a HE Pixel for January to December, 2007.

TABLE 6.4 Total Rainfall in the MPE Pixels and Mean Field Daily Bias Calculation for Year 2007

	MPE Pixel Rainfall				MPE Statistics		Rain Gauge	Month	Daily Bias		Hourly Bias		Hourly Bias Rain>0.3 mm	
	1	2	3	4	Mean	STD	Total	Bias	Mean	STD	Mean	STD	Mean	STD
	(mm)	(mm)	(mm)	(mm)	(mm)	(mm)	(mm)							
Jan	45.3	77.3	110.4	179.2	94.9	57.3	15.51	0.16	1.43	1.81	2.47	4.77	0.60	2.02
Feb	39.9	72.6	53.0	54.9	56.5	13.4	71.50	1.27	1.20	1.91	2.89	9.11	2.57	2.80
Mar	59.5	106.7	56.6	74.8	78.4	23.0	94.62	1.21	1.36	1.38	1.48	1.89	2.18	1.98
Apr	91.6	129.5	128.4	140.7	120.9	21.3	–	–	–	–	–	–	–	–
May	142.8	203.2	182.7	223.7	187.0	34.5	–	–	–	–	–	–	–	–
Jun	220.5	283.3	196.0	206.0	235.0	39.2	192.01	0.82	1.02	0.85	3.25	10.59	1.26	1.44
Jul	259.2	430.3	245.7	263.5	316.6	87.4	82.22	0.26	0.97	1.51	1.04	2.68	0.39	0.88
Aug	200.4	268.2	195.9	252.6	233.7	36.5	223.69	0.96	0.93	1.60	1.98	5.45	1.66	2.44
Sept	164.4	312.4	277.9	227.1	247.4	64.4	241.45	0.98	1.08	1.50	1.49	3.01	1.61	1.58
Oct	177.2	187.9	261.9	239.2	208.0	40.6	204.23	0.98	0.72	0.50	1.14	1.74	1.19	0.99
Nov	89.2	72.2	124.4	117.4	95.1	24.4	162.49	1.71	2.24	2.60	3.92	8.16	2.92	4.55
Dec	55.7	68.0	111.7	104.0	79.4	27.2	109.86	1.38	1.72	2.38	5.68	12.92	1.53	2.52
Year Total	1545.7	2211.4	1944.4	2083.2	1952.7	249.8	1542.3							
Avg.								0.85	1.24	1.65	2.77	8.14	1.55	2.14

Note: (–) No data values, Rain gauge total = rain gauge average for the months including the available network.

In the months of April and May some data in the rain gauge network were missing, and as a consequence, the mean field bias was calculated only for the existing data. In addition, the MPE Pixels present the complete accumulations for these months while the rain gauge column showed only the existing data. The MPE total accumulations are 120.9 and 187 mm for April and May (Table 6.4), but the MPE accumulations only for the time window that correspond to the rain gauge data are 22.41 mm and 143.61 mm for April and May, respectively and these data was not considered in the computations of Bias.

The mean field bias tended to decrease when the calculation was performed for the whole HE pixel area (16 km²). Therefore, when the MPE is accumulated (e.g., over several hours or days) the bias is reduced and the standard deviation as well. Table 6.19 provides detailed bias computations for year 2007 results.

The results indicate that the month with largest hourly bias was December (5.68), which also had the highest variability (STD =12.92). These results are decreased to 1.53 and 2.52 respectively, when the average rainfall less than 0.3 mm in radar and rain gauges were eliminated. The greatest daily

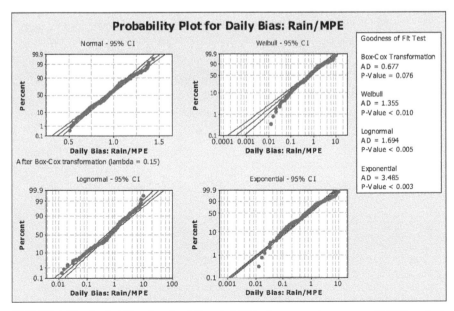

FIGURE 6.12 Probability plots for daily rainfall bias between rain gauges and MPE product.

Bias occurred in November with 2.24 and a standard deviation (STD) of 2.6. The months with Bias close to 1 are June, July, August and September but only August and September maintain the value close to one in monthly accumulations.

Different probability distributions were tested with a 95% of confidence to determine which particular distribution fits to the daily rainfall bias. The null hypothesis is that the data follow the distribution selected if P-value is greater than 0.05. The normal distribution with Box-Crox transformation ($\lambda = 0.15$) was the probability distribution that obtains a better fit to the data. Goodness of fit was evaluated using the Anderson Darling (AD) test (0.677) [3] and P-value equal to 0.677. Additionally the exponential, lognormal and Weibull distributions were tested (Figure 6.12), but obtained P-values less than 0.05 and the hypothesis was rejected, although Anderson Darling [3] values were small.

CHAPTER 7

FLOOD PREDICTION LIMITATIONS IN SMALL WATERSHEDS: PREDICTABILITY LIMITS DUE TO UP-SCALING[1, 2]

ALEJANDRA M. ROJAS-GONZÁLEZ

CONTENTS

7.1   INTRODUCTION

This Chapter analyzes the uncertainty propagation due to the model. Comparisons between rainfall resolutions and hydrologic model resolutions serve as a guide for modelers and radar developers to know how

---

[1] This chapter is an edited version from, "*Alejandra María Rojas González, 2012. Flood prediction limitations in small watersheds with mountainous terrain and high rainfall variability. Unpublished PhD Thesis for Department of Civil Engineering and Surveying, University of Puerto Rico – Mayagüez Campus*".

[2] Numbers in brackets refer to the references at the end of this book.

much detail is necessary to archive a reliable solution in small watersheds in terms of flow prediction using ensembles. This chapter presents Predictability Limits Due to Up-scaling.

7.2   PARAMETER UNCERTAINTY PROPAGATION DUE TO RAINFALL SPATIAL VARIABILITY AND HYDROLOGIC MODEL CONFIGURATIONS

Hydrologic evaluation was performed at the TBSW to evaluate the uncertainty due to spatial rainfall variations. A most comprehensive methodology was described in Section 4.4 under Chapter 4, where different interpolation methods represent rainfall coverage over MBDB model. The ensemble forecast procedure in principle draws a finite sample from the probability distribution describing the uncertainty of the initial state of the atmosphere (rainfall) or hydrologic model. Each input, parameter or model configuration combination is called the ensembles of initial condition, and each one represents a possible initial state consistent with the uncertainties in observation and analysis.

Using a deterministic model, it is possible to evaluate the propagation of the entire initial state probability distribution by the governing physical laws. The evaluation would bring information reliable to a determined initial state and would be a decision support to evaluate procedures that would be applied to obtain goodness of fit models at different resolutions or selecting a rainfall cell size when rainfall information is available at scales below NEXRAD resolutions. Here, the word "probability" is treated as conditional, because parameters were perturbed in their physical bounds, using scalar factors, selection of possible hydrologic configuration and input resolution without giving any spatial weight.

Monte-Carlo method approximation is based on a large number of possible initial hydrologic states drawn up randomly from the PDF of initial-condition uncertainty in the phase space. The stochastic dynamic simulation is constructed by a substantial amount of hydrologic simulations, repeatedly running the model is where the knowledge of the real PDF's are required. It is important that the initial ensemble member should be chosen well, their selection is further complicated by the fact that initial condition of PDF in space required for a distributed model is unknown and it changes from day to day, so that the ideal of simple random samples from this distribution cannot be achieved in practice. As a practical manner, computing time is a limiting factor at operational flood forecast centers. The modeler must make a subjective judgment balancing the number of ensemble members to include in relation to the spatial resolution of the hydrologic model used taking into consideration their physical bounds.

Using methods to resample parameters, it was possible to reduce the uncertainty due to slope degradation that result in lowest peaks and volumes retarding the runoff and smoothing the hydrograph. Five hydrologic model configurations at different scales were tested with a distributed model. The computation of the parameter statistics is shown in Table 7.1.

Grid scales are from 10 m to 400 m, with changes in total area though 3.56 km² for a high-resolution model (10 m) to 3.84 km² for coarser resolution (400 m). Average parameter values were maintained though the up-scaling at the TBSW. Terrain slope is reduced from 30.98 to 24.63% for average values and from 97 to 60.28% for maximum slopes. The most important change was due to channels cells ratio, because to increase the grid size the number of cells that represent overland and river cells are reduced. In the high-resolution model the total cells were 35,235 in which 318 cells were attributed to channel representation with a ratio of 0.9%. For coarser model resolutions up to 400 m, 18 cells were dedicated to overland process, and 6 cells for channel processes.

Additionally, rainfall and stage information are necessary to feed and validate the model. Five important events were selected from the monitoring time period (October 2007 to May 2009) for stage and rainfall. The methodology used to transform the pressure measurements of transducer installed at the outlet of the TBSW to stage measurements and posterior flow-stage curve generation has been described before in Chapter 4. Table 7.2 shows important information for the selected events, as time to peak; peak flow and average runoff depth over the TBSW. These variables compared to observed data give more descriptive information of the hydrograph shape than statistics based on error variances. The observed hydrograph for each event are displayed in Figure 7.1. The base flow was removed as a constant value from the observations because this creek has a very short concentration time due its size and high slopes.

TABLE 7.1 Descriptive Variables and Statistical Quantification for Hydrologic Model Resolution TBSW Configuration

Variable		RESOLUTION MODEL (m)				
		10	50	100	200	400
Area (km²)		3.56	3.64	3.72	3.76	3.84
Number of cells		35235	1393	342	82	18
Number of channel cells		318	61	30	12	6
Channel Cells Ratio (%)		0.90	4.38	8.77	14.63	33.33
	Minimum	0.02	0.02	0.02	0.02	0.02
Roughness	Average	0.12	0.11	0.11	0.10	0.10
	Maximum	0.15	0.15	0.15	0.15	0.15
	Minimum	27.00	10.00	10.00	0.10	1.25
Slope (%)	Average	30.98	29.83	27.69	26.21	24.63
	Maximum	97.00	87.54	86.10	70.84	60.28
	Minimum	0.15	0.64	0.64	0.64	0.64
Hyd. Conductivity (cm/h)	Average	0.69	0.69	0.69	0.69	0.70
	Maximum	2.84	0.86	0.86	0.86	0.86
Wetting Front (cm)	Average	31.62	31.62	31.62	31.62	31.62
	Minimum	0.26	0.42	0.42	0.42	0.42
Effective Porosity	Average	0.43	0.43	0.43	0.43	0.43
	Maximum	0.45	0.45	0.45	0.45	0.45
	Minimum	0	0	0	0	0
Impervious	Average	0.02	0.02	0.03	0.03	0.02
	Maximum	0.63	0.63	0.58	0.46	0.30
	Minimum	0.08	0.00	0.15	0.15	0.15
Abstraction (cm)	Average	0.80	0.80	0.78	0.80	0.84
	Maximum	1.25	1.25	1.25	1.25	1.25
Channel Width (m)	Average	5.00	5.00	5.00	5.00	5.00

Events over the year represent different initial states of the parameters and atmospheric characteristics. Antecedent soil moisture represented by initial saturation in the model is a spatially distributed parameter and it is time dependent, affecting principally the runoff depth. Low initial saturation values increase the infiltration capacity due to soil moisture and reduce the

TABLE 7.2   Inventory of Observed Events

Events	Observed peak Flow (m³/s)	Observed runoff Depth (mm)	Observed time to peak (h)
22-Oct-07	10.13	16.6	15:15
2-May-08	9.38	34.6	15:30
5-Jun-08	5.2	6.51	18:15
28-Aug-08	6.69	10.34	16:00
3-Sep-08	21.2	54.6	3:45

runoff depth. Rainfall information was collected from the rain gauge network for the events selected. Some rain gauges produced erroneous results or malfunctioned and were eliminated from the analysis. The minimum number of rain gauges used to produce a time step rainfall map was: 15 for May 2, 2008 and a maximum number of 18 rain gauges for October 22, 2007.

Table 7.3 presents storm totals for each rain gauge, average storm total for all gages and standard deviations. May 2 and September 3, 2008 events present the highest rainfall variability with a standard deviation of 24.3 mm and 20.8 mm between rain gauges; and totals rainfall of 80.4 mm and 95.7 mm respectively. Additionally, standard deviation at each rain gauge though the events were calculated at 10 min time step; presenting a maximum value of 3.29 mm, 4.29 mm, 3.23 mm, 2.88 mm and 2.59 mm for October 22, 2007; June 5, September 3, May 2, and August 28, 2008 respectively. The standard deviation calculated for both: partial and total storms reflect the spatial variability with a 4×4 km² pixel (Table 7.3).

Antecedent rainfall defines how much runoff will be produced and is an indicator of the antecedent soil moisture condition 5 days before the event occurred. The May 2 antecedent rainfall was 64.27 mm, while September 3 antecedent rainfall was only 4.41 mm. Therefore, initial soil moisture will be different for both events. Combinations of important smaller rainfall events with low and high antecedent rainfall accumulation were analyzed in this work.

Precipitation was interpolated using ArcGIS 9.3 software with the inverse distance weighted method at 10 min time steps. The method is a commonly used technique for generating weighted averaged surfaces of scatter points, and which places more weight (influence) by nearby points and less by distant points. The average storm for each event is shown in Table 7.4.

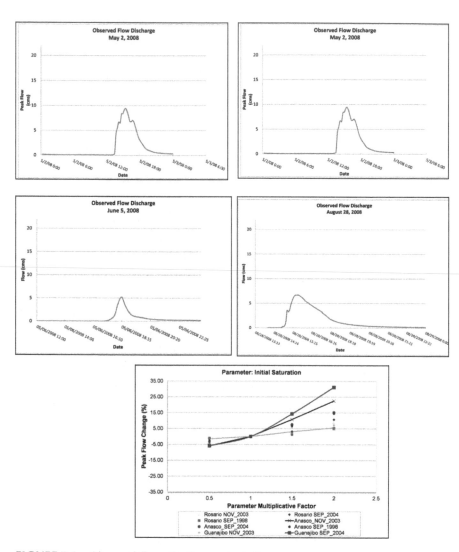

FIGURE 7.1   Observed flows for the events studied.

Convective and orographic rainfalls are the most common in western Puerto Rico and can occur daily during the wet season. In orographic events along the western coast of Puerto Rico, masses of wet air are transported by a sea breeze mechanism towards the east where it converges with the easterly trade wind over the mountains of western Puerto Rico. This, combined with the heating of the land causes the wet air to move

TABLE 7.3   Total Rainfall Event Measured in Rain Gauges Network Over 4 × 4 km² Area

Gauge station	Total Rainfall (mm)				
	22-Oct-07	2-May-08	5-Jun-08	28-Aug-08	3-Sep-08
C01	32.4	57.5	51.7	35.7	105.3
C02	38.1	–	46.7	32.6	105.4
C03	47.8	83.8	52.2	34.5	117.5
C04	40.4	86.7	51	–	–
C05	42.4	101.1	49.5	44	112.2
C06	42.7	55.7	40	31.5	–
C07	49.5	70.3	–	23.6	107.6
C08	48.6	83.3	48.8	29.9	90.2
C09	51.7	96.3	43.5	30.5	97.3
C10	43.0	94.3	–	–	–
C11	48.6	–	–	28.1	108.9
C12	45.4	82.6	34.1	14.2	97
L02	–	–	–	33.2	94.3
L03	–	40.6	–	49.9	60.3
L04	–	–	52.3	–	–
L05	32.7	–	–	11.6	38.1
L06	–	–	18.5	–	–
L07	40.1	86.8	47	37.6	82.8
L08	–	–	–	–	–
L09	–	–	44.1	49.2	116.5
L11	–	–	40.2	–	–
L13	48.5	85.3	49.5	–	100.2
L14	28.1	–	–	–	–
L15	22.5	44.9	18.6	–	–
L16	64.0	136.7	39.8	45.3	97.9
Average (mm)	42.58	80.39	42.79	33.21	95.72
STD (mm)	9.63	24.28	10.49	10.93	20.79
Antecedent rainfall: Average total rainfall previous 5 days (mm)	51.61	64.27	2.66	24.06	4.41

vertically upward forming convective cloud, within which the air is cooled and moisture is condensed causing precipitation. Convective precipitation falls over a certain area for a relative short time with a limited horizontal extent and variable intensity, forming rainfall cells over limited areas. Figure 7.2 shows the temporal variation between two selected cells after interpolation was made at 10 min time scale. Table 7.4 indicates the total storm rainfall averaged over the TBSW area, where the storm total is slightly different for each interpolation resolution.

Additionally small differences across model resolutions are due to changes in area, where the grid is intended to represent the shape of the basin.

Ogden and Julien [74] discussed the appropriateness of the correlation length as indicator of spatial structure and obtained an intergage distance of 2.5 km. Distances greater than this value will not capture the true rainfall spatial variability. With the existing average distance between the TBSW rain gauges network of 200 m, this work ensures to capture the real spatial variability for each time step though the event.

TABLE 7.4 Storm Total Produced for Different Resolutions

Model Resolution (m)	Rainfall event	Total Rain (mm)					Average (mm)	Standard deviation (mm)
		Rain grid size (meter)						
		100	200	400	1000	2000		
Grid 10	2-May-08	80.1	80.1	80.0	81.2	77.4	79.8	1.4
	3-Sep-08	100.5	100.6	100.4	97.5	101.3	100.1	1.5
	22-Oct-07	44.9	44.9	44.8	44.1	44.4	44.6	0.3
	28-Aug-08	30.2	30.3	30.3	30.2	34.6	31.1	2.0
	5-Jun-08	42.3	42.3	42.5	42.2	44.6	42.8	1.0
Grid 50	2-May-08	79.9	79.9	79.8	81.1	77.6	79.7	1.3
	3-Sep-08	100.5	100.5	100.4	97.2	101.2	100.0	1.6
	22-Oct-07	45.0	45.0	44.9	44.2	40.5	43.9	1.9
	28-Aug-08	30.0	30.0	30.0	29.8	34.4	30.9	2.0
	5-Jun-08	42.2	42.2	42.4	42.1	44.4	42.7	1.0
Grid 100	2-May-08	80.6	80.6	80.5	81.5	77.7	80.2	1.5
	3-Sep-08	100.7	100.7	100.6	98.1	101.5	100.3	1.3
	22-Oct-07	44.8	44.8	44.8	44.1	40.4	43.8	1.9
	28-Aug-08	30.8	30.8	30.8	30.8	34.9	31.6	1.8
	5-Jun-08	42.5	42.5	42.6	43.3	44.8	43.1	1.0

TABLE 7.4 Continued

Model Resolution (m)	Rainfall event	Total Rain (mm)					Average (mm)	Standard deviation (mm)
		Rain grid size (meter)						
		100	200	400	1000	2000		
Grid 200	2-May-08	80.2	79.6	79.5	80.8	76.9	79.4	1.5
	3-Sep-08	100.5	100.3	100.1	96.6	101.4	99.8	1.9
	22-Oct-07	45.0	44.7	44.6	43.9	40.2	43.7	2.0
	28-Aug-08	30.3	30.4	31.7	30.0	34.7	31.4	1.9
	5-Jun-08	42.2	42.4	42.5	42.3	44.8	42.8	1.1
Grid 400	2-May-08	78.7	79.1	80.4	80.5	77.0	79.1	1.4
	3-Sep-08	100.3	100.4	100.7	94.0	101.5	99.4	3.0
	22-Oct-07	44.7	44.9	44.7	43.5	40.3	43.6	1.9
	28-Aug-08	29.9	29.7	30.9	29.2	34.8	30.9	2.3
	5-Jun-08	44.0	42.3	42.4	42.3	44.6	43.3	1.2

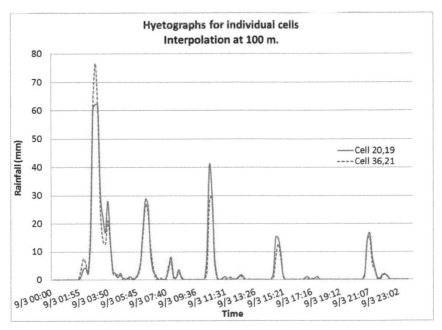

FIGURE 7.2 Hyetographs extracted from two cells (100 m resolution) for September 3, 2008.

7.3   EVALUATION OF PREDICTABILITY LIMITS

The predictability analysis due to rainfall inputs and hydrologic models resolution was performed using a total of 15,625 runs with combinations of five parameter perturbations to roughness, hydraulic conductivity and initial saturation; five hydrologic model configuration resolutions (10 m, 50 m, 100 m, 200 m, and 400 m); five rainfall resolutions (100 m, 200 m, 400 m, 1000 m and 2000 m) and five events presented in Table 7.3. The events were tested to evaluate temporal or season dependence and cover different mechanisms of rainfall generation as convective or orographic movements.

The total number of runs was reclassified in different ways depending on the type of analysis. Box plots summarize information about the shape, dispersion (confident levels of the ensemble at 5 and 95 quartiles), center of the data and outliers; also are presented as exploratory measures. A total of 125 runs that describes the dispersion of hydrologic predictions due to parameter perturbation were grouped, for each combination of model and rainfall resolution, where peak flows, runoff depth and times to peak were compared with observed data. In box plot graphs, the horizontal line represent the median of the data, the vertical lines extending from the box are called whiskers. The whiskers extend outward to indicate the lowest and highest values in the data, excluding outliers. Extreme values or outliers are represented by asterisks (*).

The event of October 22, 2007 was one of the largest flows measured at the flow gauge during the testing period, with a discharge runoff depth of 16.6 mm and peak flow of 10.11 cms, and a runoff-rainfall ratio of 0.37 (Table 7.2). October 22, 2007 ensembles show a tendency almost constant between rain resolutions, with a slight decrease of mean peak flows with increase of the rainfall resolution. Additionally, hydrologic model results are shown in the different panels for 10 m, 50 m, 100 m, 200 m and 400 m resolution (Figure 7.3A).

The averages are around the observed peak flow (red line), and hydrologic model 50 m and 100 m present outliers for high peaks in all rain gauge resolutions. In the case of runoff depth (Figure 7.6A), the average ensembles are around the observed volume (red line) with a tendency to overestimate at 10 m hydrologic model and underestimate the observed volume for the others hydrologic model resolutions in all rainfall maps. No outliers were present in runoff depth box plots. The time to peak graphs

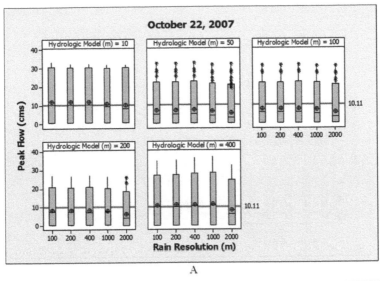

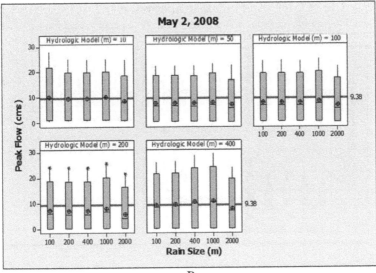

FIGURE 7.3   Box plots of peak flows for events on: (A) October 22, 2007; (B) May 2, 2008.

(Figure 7.6A) indicate low dispersions in modeled values for the 10, 100 and 200 m hydrologic models.

The event of May 2, 2008 with a discharge depth volume of 34.6 mm, and peak flow of 9.38 CMS, and a runoff-rainfall ratio of 0.43 shows a tendency almost constant for the peaks though rain sizes and hydrologic

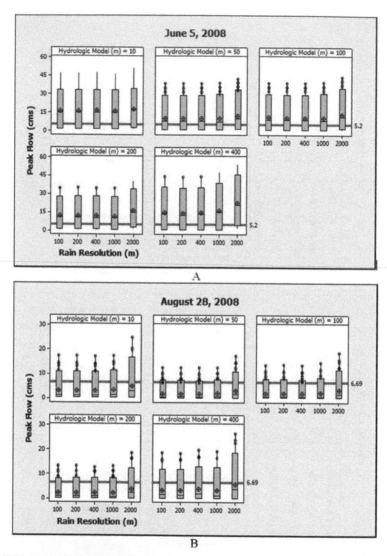

FIGURE 7.4   Box plots of Peak flows for events on: (A) June 5, 2008; (B) August 28, 2008.

models, with a slight decrease of mean peak flows with increase in the
rainfall resolution (Figure 7.3B). The average ensembles are around the
observed peak flow, and hydrologic model 200 m presents some outliers for high peaks in all rain gauges sizes. In the case of runoff depth
(Figure 7.6B), the average ensembles underestimate the runoff depth

except for the 10 m hydrologic model with 100 m rainfall size. The average ensemble for runoff depth decreases with increasing of rainfall resolution and hydrologic model resolution. No outliers were present in runoff depth box plots. Figure 7.9B shows the time to peak modeled where the average ensemble values are around the observed and low dispersions were found.

Box plots for June 5, 2008 are shown in Figure 7.5A for peak flow and Figure 7.7A for runoff depth. The event had a discharge volume of 6.51 mm and 5.2 cms flow, and a runoff-rainfall ratio of 0.154.

The average ensembles tended to overestimate peaks and volumes as well, therefore, showing a tendency almost constant for the peak average though rain sizes and hydrologic models, with an increase of mean peak flows with increase rainfall resolution (Figure 7.4A) for the 400 m hydrologic model. Hydrologic models presented some outliers for high peaks in all rain gauges sizes, except for the 10 m hydrologic model. In the case of runoff depth (Figure 7.7A), the simulations for 10 m resolution model were out of the observed volume and the others ones ensembles slightly covering the observed volume. The average resemble of runoff

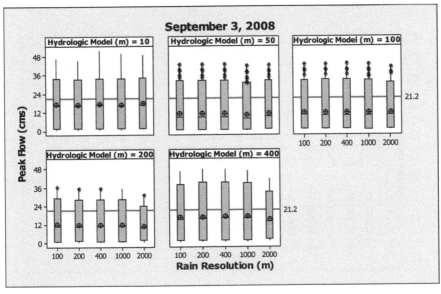

FIGURE 7.5 Box plots of Peak flows for September 3, 2008 event.

depth decrease to increase the rainfall resolutions and hydrologic model resolution. No outliers are presented in runoff depth box plots. Time to peak ensemble means (Figure 7.10A) are within the observed value of 18:15 min for June 5, 2008 with underestimation in hydrologic models greater than 50 m. For hydrologic models 200 and 400 m the quartile 95 are below the observed value.

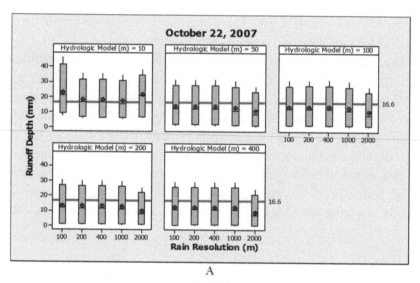

A

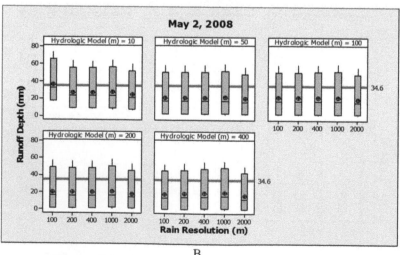

B

FIGURE 7.6 Box plots for runoff depth: (A) October 22, 2007; (B) May 2, 2008.

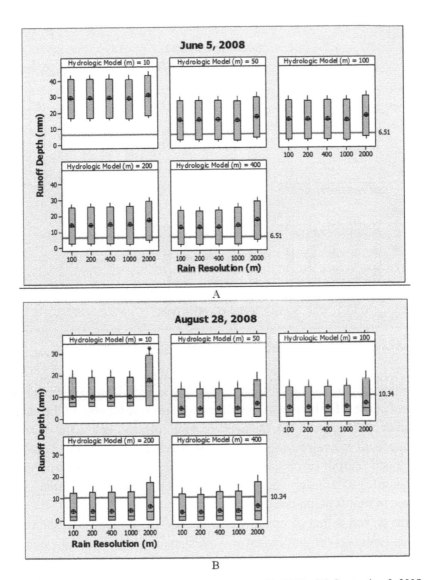

FIGURE 7.7   Box plots for runoff depth: (A) August 28, 2008; (B) September 3, 2008.

The event of August 28, 2008 has a discharge depth volume of 10.34 mm, 6.69 cms peak flow, and a runoff-rainfall ratio of 0.34. It shows a tendency almost constant between rain sizes, with a slighter increase of mean peak flows with increase of the rainfall resolution, additionally the range between quartiles 5 and 95 is also increased (Figure 7.4B). The average

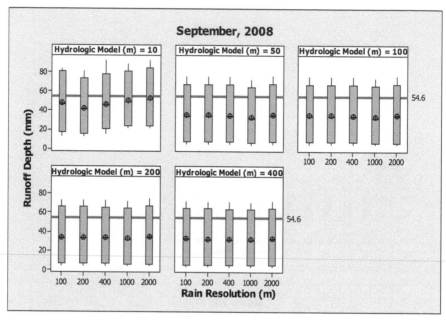

FIGURE 7.8   Box plots for runoff depth for September 3, 2008.

ensembles are below the observed peak flows, and all hydrologic models present outliers for high peaks in all rain gauges resolutions. In the case of runoff depth (Figure 7.7B), the average ensembles are below the observed volume with a tendency to underestimate, except for 10 m hydrologic model and rainfall resolution of 2000 m. Therefore, for some ensembles the quartiles 95 are very close to the observed volume. No outliers were present in runoff depth box plots.

The reason is that computations with very low initial saturation (0.25) did not represent the antecedent soil moisture and high hydraulic conductivities. Figure 7.10B shows the time to peak box plots showing values around the observed (August 28 16:00) with low dispersion for the hydrologic model of 100 m resolution. The hydrologic models with more dispersion are 50 and 400 m resolution.

The event of September 3, 2008 was the largest peak flow measured at the flow gauge in the studied period, with a discharge depth volume of 54.6 mm, 21.2 cms peak flow, and a runoff-rainfall ratio of 0.5. September 3, 2008 shows a tendency almost constant between rain sizes, with slight

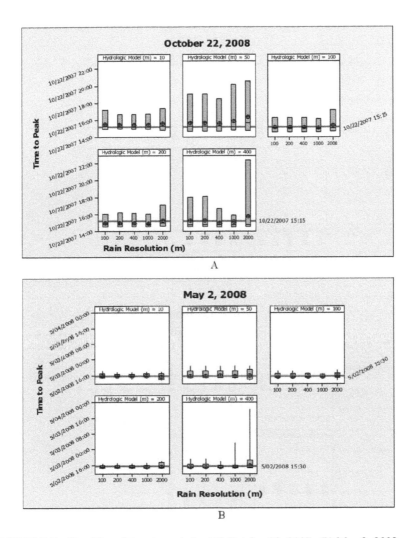

FIGURE 7.9   Box Plot of time to peak for (A) October 22, 2007; (B) May 2, 2008.

changes of mean ensemble peak flows (Figure 7.5). The ensemble averages are underestimating the observed peak flow and the 10 m hydrologic
model results are closer to the observed values as is the 400 m resolution
hydrologic model as well. Hydrologic model at 50 m, 100 m and 200 m
present outliers for high peaks in all rain gauge resolutions. In the case of
runoff depth (Figure 7.8), the average ensembles at 10 m hydrologic model

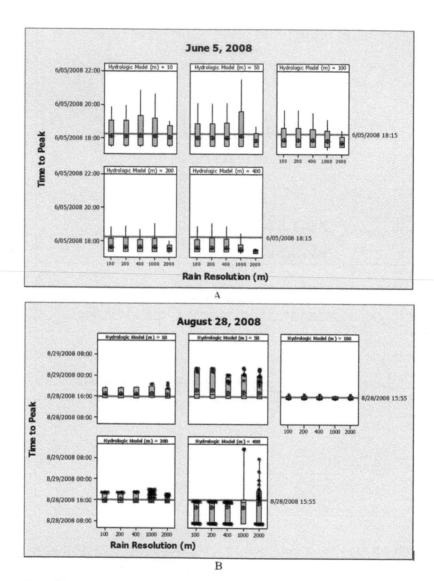

FIGURE 7.10   Box Plot of time to peak for (A) June 5, 2008; (B) August 28, 2008.

are around the observed depth volume with a tendency to underestimate the observed depth volume. The observed runoff depth volume is near to the quartile 95 for 50, 100, 200 and 400 m hydrologic model resolutions. No outliers were present in volume depth runoff box plots.

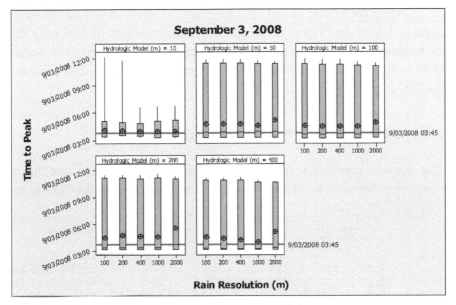

FIGURE 7.11   Box Plot of time to peak for September 3, 2008.

Figure 7.11 indicated high dispersions for the 50, 100, 200 and 400 m hydrologic models resolutions and a tendency to overestimate the observed time to peak (September 3, was 3:35). The significant dispersions are due to the form of the observed hydrograph that consist in three limbs. With low initial saturations and high hydraulic conductivities the first jump is absorbed and peaks are greater in the second or third limb.

In general the average ensembles were underestimating the peak flow and runoff depth for the analyzed events, except for June 5, 2008 where the contrary situation was obtained. This event is characterized by an antecedent dry period and medium rainfall in a short time, revealing an anomaly for dry periods and lighter rainfall events.

7.4   EVALUATING HYDROLOGIC MODELS RESOLUTIONS AND RAINFALL RESOLUTIONS: PROBABILITY PLOTS

The 15,625 runs were grouped in a different way that helps to explain differences between rainfall resolutions and hydrologic model resolutions as well. Probability with normal distribution and confident levels

(5–95) were calculated and plotted for ensemble with observed values in Figures 7.12–7.16. The ensembles for example consist of 625 runs for each hydrologic model including the perturbation parameters and variations in rainfall sizes. Goodness of fit statistics was calculated to compare the data to probability distribution.

7.4.1 PEARSON CORRELATION COEFFICIENT

The Pearson correlation coefficient measures the strength of the linear relationship between the X and Y variables on a probability plot (The value

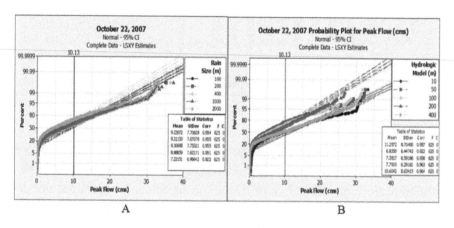

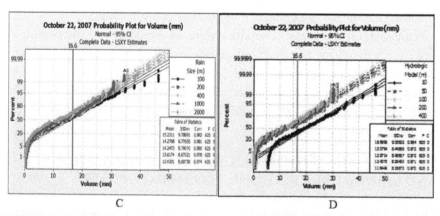

FIGURE 7.12 Probability plots for October 22, 2007: (A) Rain ensembles for peak flow, (B) Hydrologic model ensembles for peak flow, (C) Rain Ensembles for discharge depth volume, (D) Hydrologic Model ensembles for discharge depth volume.

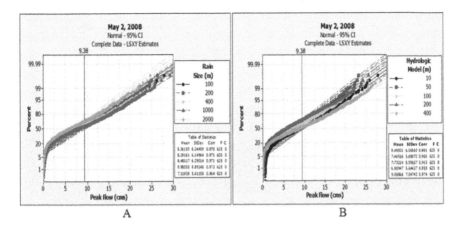

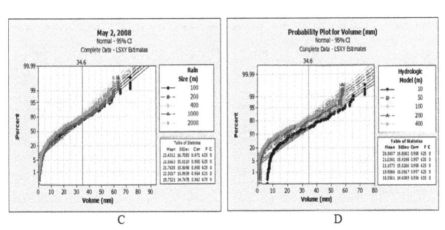

FIGURE 7.13 Probability plots for May 2, 2008: (A) Rain ensembles for peak flow, (B) Hydrologic model ensembles for peak flow, (C) Rain Ensembles for discharge depth volume, (D) Hydrologic Model ensembles for discharge depth volume.

close to 1 indicates that the relationship is highly linear). Almost all graphs present Pearson correlation coefficient values above 0.93. The event that presents the lowest was August 28, 2008 (Figure 7.15A) for peak flows with 0.875 coefficient of determination. Additional information such as mean and standard deviation of the ensemble are shown in Figures 7.12–7.16. The lowest extreme values in peak and runoff depth did not have good agreement with the PDF, and was produced by low initial soil saturation values (0.25) in combination with high hydraulic conductivities. In general, the ensemble means and standard deviation decreased with increasing rain resolution input or increase of model resolution.

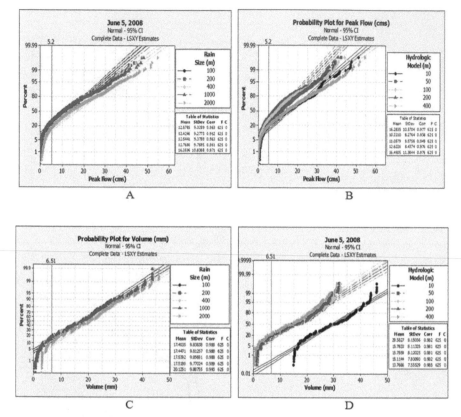

FIGURE 7.14 Probability plots for June 5, 2008: (A) Rain ensembles for peak flow, (B) Hydrologic model ensembles for peak flow, (C) Rain Ensembles for discharge depth volume, (D) Hydrologic Model ensembles for discharge depth volume.

7.4.2 RPS COMPUTATION

The statistical measures (Bias, MSE, RMSE and the RPS) were calculated for the 625 members for each ensemble explained above. The RPS compares each category with observed values; 12 categories were selected for the RPS computation. Table 7.5 shows the statistics calculated for October 22, 2007 where the lowest RPS for peak flow variable and different rainfall resolutions are for rainfalls of 100 m (0.79) and 400 m (0.79) with similar RMSE (8.09 mm and 8.05 mm, respectively); and 100 m (RPS: 0.7) follow by 200 m (RPS: 0.77) and 400 m (RPS: 0.77) for runoff depth. Therefore, the lowest

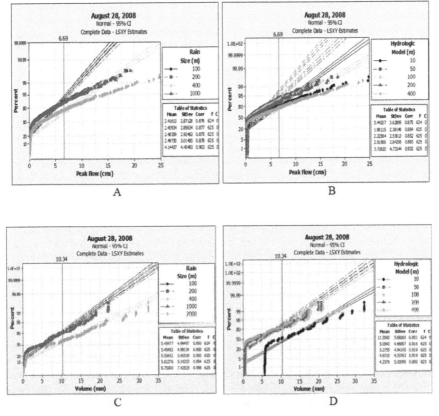

FIGURE 7.15 Probability plots for August 28, 2008: (A) Rain ensembles for peak flow, (B) Hydrologic model ensembles for peak flow, (C) Rain Ensembles for discharge depth volume, (D) Hydrologic Model ensembles for discharge depth volume.

RMSE (9.23 mm) is for 400 m rainfall resolution. The time to peak presents the best RPS (0.43) for 400 m rainfall with the lowest RMSE (49 min) and the Bias is close to one. When the ensembles grouped by hydrologic model were analyzed, the best RPS for peak flow are 0.78 and 0.79 for the 400 m and 200 m hydrologic models, respectively. The best lowest RMSE, 6.91 cms is for 400 m and 7.52 cms for 200 m. Analyzing the runoff depth volume variable, the 10 m hydrologic model obtained a good RPS (0.75) as did the 50 m (RPS: 0.83) and 100 m (RPS: 0.83) m hydrologic model.

The hydrologic model that produced the best time to peak according to the RPS is 10 and 100 m resolution models with 0.38 and 0.41 RPS's,

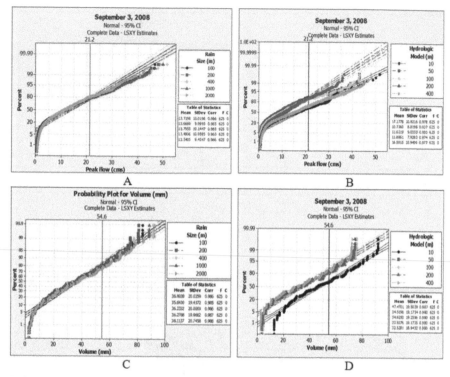

FIGURE 7.16 Probability plots for September 3, 2008: (A) Rain ensembles for peak flow, (B) Hydrologic model ensembles for peak flow, (C) Rain Ensembles for discharge depth volume, (D) Hydrologic Model ensembles for discharge depth volume.

additionally, these resolutions present lower RMSE's of 40 and 34 min. Table 7.6 presents the ensemble statistics and skill of the prediction according to rainfall resolution and hydrologic model resolution for the event occurring on May 2, 2008. Evaluating peak flow and time to peak due to rainfall variations the RPS's do not clearly favor any resolutions. Therefore 100 m, 200 m and 400 m resolution obtain similar value of RPS. In the case of runoff depth volume, the RPS favors rainfall resolutions of 100 m and 1000 m with RPS values of 1.28 and 1.36, respectively.

Ensembles grouped by hydrologic resolution provide RPS values that favor the 10 and 100 m resolution for peak flow, volume and time to peak.

Table 7.7 shows the statistics and skills of the prediction for June 5, 2008 where the rainfall resolutions favor the 100 m, 200 m and 400 m

TABLE 7.5 Ensemble Statistics and Skill of Prediction According to Rainfall Resolution and Hydrologic Model Resolution for October 22, 2007

	Rainfall					Hydrologic Model				
	100	200	400	1000	2000	10	50	100	200	400
Peak										
Bias	0.91	0.91	0.91	0.92	0.88	0.71	1.21	0.67	0.73	0.77
MAE	6.74	6.74	6.73	6.78	6.79	6.99	8.36	6.64	6.40	5.93
RMSE	8.09	8.09	8.05	8.13	8.08	8.09	10.16	7.71	7.52	6.91
RPS	**0.79**	0.80	**0.79**	0.83	1.08	0.86	1.10	0.95	0.79	**0.78**
Volume										
Bias	1.50	1.41	1.41	1.34	1.23	2.05	1.29	1.28	1.22	1.15
MAE	8.32	7.92	7.94	8.05	8.24	8.99	8.05	8.07	8.12	8.32
RMSE	10.02	9.21	9.23	9.32	9.81	11.21	9.35	9.38	9.48	9.74
RPS	**0.70**	**0.77**	**0.77**	0.85	0.88	**0.75**	0.83	0.83	0.94	1.03
Time										
Bias	1.002	1.002	0.998	1.002	1.026	1.014	1.036	0.993	0.986	1.001
MAE	0:34	0:34	0:29	0:35	0:50	0:24	0:51	0:25	0:29	0:54
RMSE	1:09	1:12	0:49	1:17	1:45	0:40	1:40	0:34	0:35	2:04
RPS	0.47	0.45	**0.43**	0.47	0.72	**0.38**	0.69	0.41	0.66	0.94

hydrologic model for peak flow, depth volume and time to peak with the lowest RPS values around 0.6, 1.7 and 0.77, respectively. Therefore, the RMSE are very similar between the resolutions. There was no clarity in terms of the best hydrologic model resolution, because the peak flow favored the 50 and 100 m resolution; runoff depth volume favored the 200 and 400 m and time to peak favored the 10 and 50 m resolution model, respectively.

The August 28, 2008 event in Table 7.8 indicates the statistics and skills of the ensembles where the RPS favored the rainfall resolution of 2000 m with 0.76 for peak flow and 1.11 for depth volume. Time to peak did not present differences between 100 m, 200 m and 400 m rainfall resolution. The skill ensemble by hydrologic models gave the lowest RPS for 50 m resolution for peak flow and the second lowest value for time to peak.

The mean ensemble for peak and volume are underestimated for the event occurring on September 3, 2008, where the RPS are similar between

TABLE 7.6 Ensemble Statistics and Skill of the Prediction According to Rainfall Resolution and Hydrologic Model for May 2, 2008

	Rainfall					Hydrologic Model				
	100	200	400	1000	2000	10	50	100	200	400
Peak										
Bias	0.89	0.88	0.90	**0.96**	0.76	1.01	0.80	0.82	0.74	1.02
MAE	5.57	5.51	5.58	5.75	5.38	5.20	5.38	5.58	5.48	6.15
RMSE	6.49	6.40	6.51	6.72	6.22	6.16	6.18	6.41	6.34	7.21
RPS	0.73	0.73	0.73	0.76	0.75	**0.71**	**0.75**	0.78	0.79	0.80
Volume										
Bias	2.50	2.30	2.32	2.38	2.11	3.02	2.24	2.26	2.12	1.96
MAE	17.63	18.01	17.94	17.95	18.67	15.23	18.31	18.31	18.79	19.55
RMSE	20.45	20.63	20.56	20.59	21.30	17.42	21.01	21.01	21.50	22.27
RPS	**1.28**	1.45	1.46	**1.36**	1.69	**0.95**	**1.49**	**1.47**	1.64	1.84
Time										
Bias	0.992	0.992	0.990	0.991	1.003	1.001	1.014	0.991	0.981	0.985
MAE	0:30	0:29	0:29	0:32	0:46	0:22	0:41	0:27	0:29	0:46
RMSE	0:44	0:42	0:41	0:57	1:55	0:33	1:09	0:36	0:34	1:56
RPS	0.20	**0.18**	**0.18**	0.21	0.27	**0.14**	0.30	**0.15**	0.22	0.55

rainfall resolutions for peak flow and the 10 m and 400 m hydrologic model are favored. The depth volume variable and the time to peak favored the rainfall resolution of 2000 m and a hydrologic model of 10 m followed by 50 m.

Beven [14] has recognized that the nonuniqueness of a model, especially in distributed models similar to the one used in this research, can produce results close to the observed peak flow, runoff depth and time to peak, using different combination of parameters and inputs. Results in this chapter also reveal the coexistence of alternative parameter sets that provide a suitable framework for model calibration and uncertainty estimation. The configuration ensemble that was out of the range around the peak flow, 5 and 95 quartiles and minimum peak flow estimation, was the model at 10 m resolution with all rainfall resolutions. This ensemble overestimates simulated flows and cannot reproduce flows for June 5, 2008. For the time to peak the ensembles for hydrologic model 100 m and

TABLE 7.7   Ensemble Statistics and Skill of the Prediction According to Rainfall Resolution and Hydrologic Model for June 5, 2008

	Rainfall					Hydrologic Model				
	100	**200**	**400**	**1000**	**2000**	**10**	**50**	**100**	**200**	**400**
Peak										
Bias	2.42	2.39	2.41	2.45	3.14	3.13	1.96	2.13	2.42	3.16
MAE	8.86	8.75	8.87	9.19	11.87	12.02	6.95	7.58	8.61	12.38
RMSE	12.14	12.01	12.16	12.63	15.74	15.45	10.11	10.75	11.36	16.17
RPS	**0.61**	**0.60**	**0.61**	0.63	0.99	1.06	**0.49**	**0.58**	0.62	1.19
Volume										
Bias	3.34	3.35	3.37	3.36	3.86	5.68	3.03	3.03	2.90	2.64
MAE	11.90	11.92	12.01	12.01	14.06	23.07	10.22	10.23	9.69	8.69
RMSE	14.73	14.74	14.84	14.76	16.86	24.51	12.40	12.41	11.71	10.53
RPS	**1.70**	**1.71**	**1.72**	1.78	2.26	4.86	1.50	1.49	**1.43**	**1.15**
Time										
Bias	0.978	0.978	0.978	0.978	0.971	0.991	0.984	0.976	0.966	0.963
MAE	0:30	0:29	0:30	0:32	0:33	0:23	0:25	0:28	0:37	0:41
RMSE	0:33	0:33	0:34	0:37	0:36	0:29	0:32	0:31	0:39	0:42
RPS	**0.78**	**0.77**	**0.77**	0.82	1.09	**0.36**	**0.53**	0.82	1.30	1.62

2000 m rainfall; 200 m hydrologic model and rains: 400 m, 1000 m and 2000 m; 400 m hydrologic with rains: 100 m, 200 m, 1000 m 2000 m are out of 95% confident level. June 5, 2008 is characterized by dry conditions and low peak flow and volume.

The ensembles that can reproduce well the time to peak when the hydrographs present 2 limbs (October 22, 2007); or 3 bumps (September 3, 2008) are the 10 m hydrologic model for all rainfall resolution for September and the 10 m, 100 m and 200 m hydrologic models for all rainfall resolutions. For events with only one limb like August 28 and May 2, 2008 the best models with low dispersions around the observed time to peak were 10 m, 100 m and 200 m hydrologic models.

Based on the RPS calculated for the rainfall resolution ensembles in combination with all models resolution (625 members for each event) and parameter perturbations the best rainfalls simulations were observed at the 100 m for peak flow followed by 200 m and 400 m with RPS values very

TABLE 7.8   Ensemble Statistics and Skill of the Prediction According to Rainfall Resolution and Hydrologic Model for August 28, 2008

	Rainfall					Hydrologic Model				
	100	200	400	1000	2000	10	50	100	200	400
Peak										
Bias	0.36	0.36	0.37	0.38	0.62	0.51	0.30	0.33	0.38	0.58
MAE	4.94	4.93	4.92	4.94	4.79	4.74	5.12	4.99	4.82	4.87
RMSE	5.34	5.34	5.33	5.36	5.50	5.25	5.47	5.36	5.26	5.54
RPS	1.07	1.07	1.06	1.05	0.76	0.87	1.20	1.13	1.02	0.78
Volume										
Bias	0.82	0.82	0.83	0.84	1.31	1.73	0.76	0.79	0.70	0.65
MAE	8.43	8.42	8.37	8.30	7.72	5.46	8.77	8.65	9.04	9.32
RMSE	9.41	9.41	9.36	9.34	8.93	6.35	9.72	9.62	9.97	10.25
RPS	1.73	1.73	1.68	1.67	**1.11**	**0.66**	1.95	1.91	1.96	2.07
Time										
Bias	1.007	1.007	0.997	0.998	1.005	1.058	1.114	0.973	1.012	0.858
MAE	1:31	1:30	1:16	1:33	1:08	1:03	2:03	0:31	0:46	2:34
RMSE	2:47	2:46	2:19	3:01	2:18	1:30	3:27	0:36	1:07	4:24
RPS	0.36	0.36	0.38	0.37	0.40	0.21	0.49	0.57	0.42	0.92

similar. For runoff depth the rainfall at 100 m gives the better RPS for 3 events and the exceptions favor 2000 m for August 28 and September 3, 2008 (Table 7.9). The RPS for time to peak favored 200 m followed by 400 m rainfall resolution. These findings reveal that the hypothesis that the 100 m rainfall resolution will produce the best ensemble behavior for any event is rejected. The rainfall quantification due to rainfall interpolation will produce similar hydrologic ensembles behavior.

In the case of the hydrologic model resolution, the hypothesis formulated was that the hydrologic model with high-resolution (10 m) will generate the best ensemble behavior for the events analyzed. This statement is true only for 2 events evaluating the peak flow variable. For runoff depth, the 10 m hydrologic model did not produce the best RPS for dry conditions and light rainfall event (June 5, 2008) with a storm total rainfall of 42.79 mm. the high-resolution model obtained the better behavior for time to peak. This resolution model is not operationally practical for

TABLE 7.9 Ensemble Statistics and Skill of the Prediction According to Rainfall Resolution and Hydrologic Model for September 3, 2008

	Rainfall					Hydrologic Model				
	100	200	400	1000	2000	10	50	100	200	400
Peak										
Average	13.72	13.67	13.80	13.48	13.34	17.18	10.74	11.62	11.89	16.58
Bias	0.65	0.64	0.65	0.64	0.63	0.81	0.51	0.55	0.56	0.78
MAE	11.13	11.15	11.18	11.36	10.95	10.03	12.75	12.01	10.82	10.17
RMSE	12.75	12.77	12.82	12.95	12.50	11.72	14.04	13.46	12.34	12.07
RPS	1.06	1.07	1.07	1.11	1.08	0.78	1.54	1.35	1.24	0.77
Volume										
Average	36.80	35.64	36.23	36.28	38.11	47.47	34.52	34.62	33.93	32.53
Bias	1.74	1.68	1.71	1.71	1.80	2.24	1.63	1.63	1.60	1.53
MAE	22.40	22.85	22.90	22.80	22.14	17.75	23.47	23.45	23.80	24.62
RMSE	26.94	27.31	27.30	27.23	26.63	20.93	27.97	27.96	28.40	29.29
RPS	1.20	1.28	1.26	1.25	**1.13**	**0.79**	1.35	1.33	1.37	1.54
Time										
Average	4:30	4:29	4:25	4:20	4:58	3:58	4:47	4:37	4:49	4:30
Bias	1.20	1.20	1.18	1.16	1.32	1.06	1.28	1.23	1.28	1.20
MAE	0:30	0:29	0:29	0:32	0:46	0:22	0:41	0:27	0:29	0:46
RMSE	0:44	0:42	0:41	0:57	1:55	0:33	1:09	0:36	0:34	1:56
RPS	0.54	0.52	0.51	**0.48**	0.50	**0.22**	**0.32**	0.45	0.88	0.92

larger basins, and therefore an alternative has to be selected. The RPS analysis favored the 200 m model resolution for time to peak (5 events), runoff depth (4 events) and peak flow (3 events) followed by 400 m model resolution principally for peak flow.

7.5 SELECTION OF THE OPTIMAL RAINFALL AND GRID RESOLUTION FOR THE MBDB MODEL

The goal of the research study in Part 1 was to develop recommendations for rain and grid resolutions that will provide equal accuracy with a 100 m and 10 m rainfall and grid resolution model, respectively (i.e., the smallest

TABLE 7.10   Mean RPS Values for Peak Flow, Volume and Time to Peak for 5 Storms, 5 Rainfall Resolutions and 5 Grid Resolutions

Storm	Rainfall Resolution					Grid Resolution				
	100 m	200 m	400 m	1000 m	2000 m	10 m	50 m	100 m	200 m	400 m
	Peak Flow RPS					Peak Flow RPS				
3-Sep-2008	1.06	1.07	1.07	1.11	1.08	0.78	1.54	1.35	1.24	0.77
5-Jun-2008	0.61	0.60	0.61	0.63	0.99	1.06	0.49	0.58	0.62	1.19
28-Aug-2008	1.07	1.07	1.06	1.05	0.76	0.87	1.20	1.13	1.02	0.78
22-Oct-2008	0.79	0.80	0.79	0.83	1.08	0.86	1.10	0.95	0.79	0.78
2-May-2008	0.73	0.73	0.73	0.76	0.75	0.71	0.75	0.78	0.79	0.80
MEAN	**0.85**	**0.85**	**0.85**	**0.88**	**0.93**	**0.86**	**1.02**	**0.96**	**0.89**	**0.86**
	Runoff Depth RPS					Runoff Depth RPS				
3-Sep-2008	1.20	1.28	1.26	1.25	1.13	-	1.35	1.33	1.37	1.54
5-Jun-2008	1.70	1.71	1.72	1.78	2.26	-	1.50	1.49	1.43	1.15
28-Aug-2008	1.73	1.73	1.68	1.67	1.11	-	1.95	1.91	1.96	2.07
22-Oct-2008	0.70	0.77	0.77	0.85	0.88	-	0.83	0.83	0.94	1.03
2-May-2008	1.28	1.45	1.46	1.36	1.69	-	1.49	1.47	1.64	1.84
MEAN	**1.32**	**1.39**	**1.38**	**1.38**	**1.41**	-	**1.42**	**1.41**	**1.47**	**1.52**
	Time to Peak RPS					Time to Peak RPS				
3-Sep-2008	0.54	0.52	0.51	0.48	0.50	0.22	0.32	0.45	0.88	0.92
5-Jun-2008	0.78	0.77	0.77	0.82	1.09	0.36	0.53	0.82	1.30	1.62
28-Aug-2008	0.36	0.36	0.38	0.37	0.40	0.21	0.49	0.57	0.42	0.92
22-Oct-2008	0.47	0.45	0.43	0.47	0.72	0.38	0.69	0.41	0.66	0.94
2-May-2008	0.20	0.18	0.18	0.21	0.27	0.14	0.30	0.15	0.22	0.55
Mean	**0.47**	**0.46**	**0.45**	**0.47**	**0.59**	**0.26**	**0.46**	**0.48**	**0.70**	**0.99**

resolutions evaluated). To achieve this objective, the RPS values are summarized in Table 7.10; and were evaluated in a Two-Way ANOVA test. The RPS data were determined to be normally distributed and have equal variances, which is a requirement for the Two-Way ANOVA test.

The goal of the evaluation is to determine significant differences between the mean of the RPS for the highest resolution (100 m rainfall resolution and 10 m grid resolution) and the means for the other resolutions. If there is no significant difference between the mean of the RSP for the finer resolution and a coarser resolution, then the model can be up-scaled to the coarser resolution without loss of accuracy relative to the finer resolution. A gray highlighted cell in Table 7.10 indicates that a significant difference exists between that resolution and the highest resolution. For rainfall resolution, there is a significant difference between the 100 m resolution and the 2000 m resolution. For the grid resolution, there is a significant difference between the 10 m resolution and the 200 and 400 m resolutions. Therefore, based on the Two-Way ANOVA analysis of the RPS, the recommended up-scaled rainfall resolution, which will provide equivalent accuracy with the 100 m rainfall resolution, is 1000 m, and the recommended up-scaled grid resolution, which will provide equivalent accuracy with the 10 m resolution, is 100 m.

CHAPTER 8

FLOOD PREDICTION LIMITATIONS IN SMALL WATERSHEDS: CALIBRATION/VALIDATION OF A DISTRIBUTED HYDROLOGIC MODEL AT MBDB [1, 2]

ALEJANDRA M. ROJAS-GONZÁLEZ

CONTENTS

8.1 INTRODUCTION

This chapter reveals findings in Chapters 1–7 applied to the MBDB using a distributed model with a resolution of 200 m and radar information for 2003. Predictability limits (maximum and minimum peak flows

---

[1] This chapter is an edited version from, *"Alejandra María Rojas González, 2012. Flood prediction limitations in small watersheds with mountainous terrain and high rainfall variability. Unpublished PhD Thesis for Department of Civil Engineering and Surveying, University of Puerto Rico – Mayagüez Campus".*

[2] Numbers in brackets refer to the references at the end of this book.

and runoff depths) were calculated for the calibration developed at the basins. The hydrologic model of 100 m was recommended in the previous section. However, the 200 m hydrologic model was tested because not significance differences were found for peak flow and runoff depth, variables analyzed here.

8.2 CALIBRATION/VALIDATION OF A DISTRIBUTED HYDROLOGIC MODEL AT MBDB

The rainfall source used to run one-year simulation (2003) was the NWS MPE radar-rainfall products. This source has a mean systematic error (Bias) correction for Puerto Rico and in some places cannot remove the local bias, correctly, principally for small areas. In section 6 of this chapter, an evaluation of the efficiency in removing the local Bias from MPE was conducted at the TBSW and additionally bias corrections need to be developed for small subwatersheds.

At observed flow locations, the base flow must be removed to obtain runoff observations. The PART computer program analyzes daily streamflow records and estimates a daily ground water discharge. The method designates groundwater discharge to be equal to streamflow on days that fit a requirement of antecedent recession; linearly interpolates groundwater discharge for other days; and is applied to a long period of record to obtain an estimate of the mean rate of groundwater discharge and remove base flow at daily a time step [83].

8.2.1 MONTHLY BASE FLOW SEPARATION

Table 8.1 shows the results for monthly base flow separation for 2003 at three USGS stream flow stations obtained from the PART computer model (Figure 8.1A–8.1C for Río Guanajibo near Hormigueros, Río Grande de Añasco near San Sebastian and Río Rosario near Hormigueros, respectively). Additionally daily computations were obtained to add them to the Vflo runoff results for comparison with the observed stream flow.

Figure 8.2 shows the simulated and observed accumulated runoff depth for the three USGS stations for 2003. The percent of errors for runoff

TABLE 8.1 Base Flow Separation at Three USGS Streamflow Stations for 2003

	Guanajibo near Hormigueros			Añasco near San Sebastian			Rosario near Hormigueros		
	Stream flow	Base flow	Runoff	Stream flow	Base flow	Runoff	Stream flow	Base flow	Runoff
	mm	mm	mm	mm	mm	mm	mm	mm	mm
Jan	10.2	9.4	0.8	44.2	41.9	2.3	32.0	30.7	1.3
Feb	4.8	4.1	0.8	28.2	25.7	2.5	25.1	23.6	1.5
March	4.1	2.8	1.3	20.6	18.5	2.0	19.3	17.3	2.0
April	33.3	14.7	18.5	59.9	30.0	30.0	45.0	26.7	18.3
May	28.2	18.5	9.7	231.6	139.4	92.2	97.3	70.4	26.9
Jun	8.9	7.6	1.3	90.4	70.6	19.8	66.5	48.5	18.0
Jul	9.7	6.9	2.8	46.5	37.8	8.6	57.7	41.4	16.3
Aug	12.2	8.4	3.8	97.3	53.3	43.9	59.9	42.7	17.3
Sep	45.7	25.4	20.3	136.7	68.6	68.1	99.1	61.5	37.6
Oct	123.4	76.7	46.7	280.7	142.7	137.9	234.4	167.4	67.1
Nov	235.0	122.9	112.0	454.2	255.5	198.6	265.4	170.4	95.0
Dec	72.6	48.5	24.1	170.2	125.5	44.7	122.4	94.5	27.9
Total	**588.0**	**345.9**	**242.1**	**1660.4**	**1009.7**	**650.7**	**1124.2**	**795.0**	**329.2**

depth around these values were 1.81%, 1.07% and 4.47% for Guanajibo, Añasco and Rosario USGS outlet points. Nash–Sutcliffe model efficiency coefficients calculated for these outlet points were 0.46, 0.10 and 0.02, respectively.

Some systematic errors in the MPE rainfall product were revealed in the simulation period, where the MPE sensed larger amounts of rainfall than actually occurred within the study MBDB area. In this cases the observed discharges were lower than the simulated (Figures 8.3A and 8.3B) for Añasco and Rosario rivers. Additionally, maximum and minimum discharges were calculated perturbing the roughness and hydraulic conductivity within their limits evaluated in previous sections (0.25 and 1.75, respectively), while setting the initial saturation to 0.25 and 0.95, respectively. It is clear that, for certain rainfall events, large differences between the modeled and observed data exist (Figures 8.4A and 8.4B),

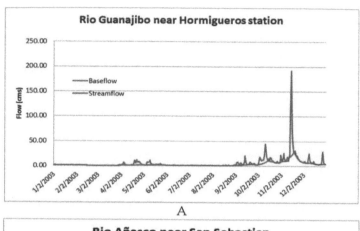

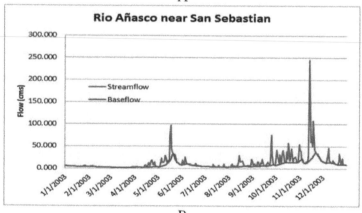

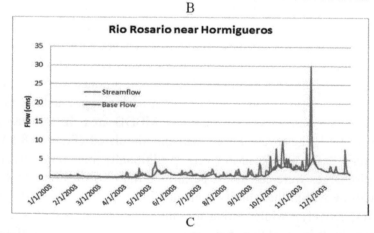

FIGURE 8.1   Daily stream flow and baseflow computation for 3 USGS stations, 2003.

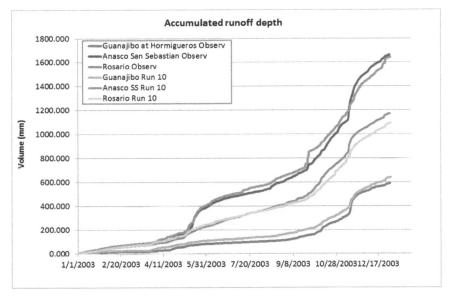

FIGURE 8.2 Runoff depth accumulated for the USGS stations for 2003.

indicating systematic errors due to MPE rainfall quantification, and limiting flood predictability in western Puerto Rico using the MPE radar product.

The stream flow examples in Figures 8.4A and 8.4B, illustrate cases in which the up-scaled model could not reproduce the observed flow because the rainfall could not be quantified accurately using the MPE product. Forcing the model to produce maximum and minimum peak flows by judiciously parameterizing the model showed that the predictability limits of the model were well above the magnitude of the observed flow.

8.3 SUMMARY

The implications of these results are that a better rainfall product is needed within the study area before accurate flood forecasts can be expected. It is hoped that the high-resolution CASA radar product, currently under development, may fulfill this important need.

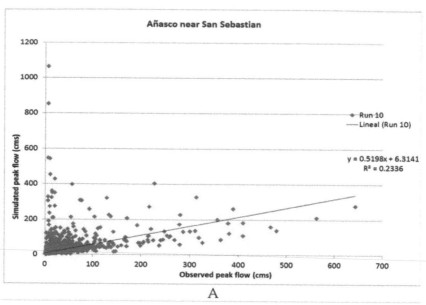

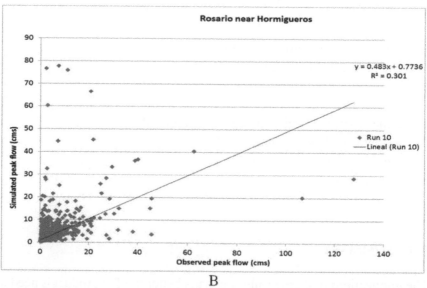

FIGURE 8.3   Comparison between observed and simulated discharge for 2003 at hourly time step for: (A) Río Grande de Añasco station near San Sebastian, Y = (0.54X + 0.314), $R^2 = 0.254$ and (B) Río Rosario station near Hormigueros, Y = (0.48X + 0.774), $R^2 = 0.301$.

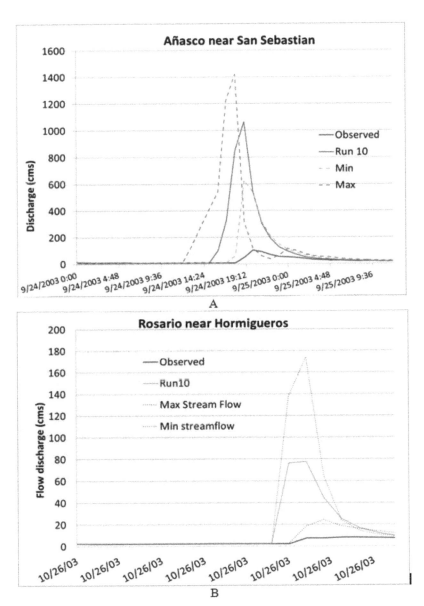

FIGURE 8.4   Maximum, minimum and observed runoff for Añasco river (A) and Rosario river (B) outlet points for selected events.

CHAPTER 9

FLOOD PREDICTION LIMITATIONS IN SMALL WATERSHEDS: CONCLUDING REMARKS[1, 2]

ALEJANDRA M. ROJAS-GONZÁLEZ

CONTENTS

9.1 INTRODUCTION

A hydrologic model was evaluated for its potential to perform real-time flood forecasting within the Mayagüez Bay drainage basin (MBSB, 819.1 km²), located in western Puerto Rico. Minimal run times, enhanced prediction skill, parameterization of variables and the understanding the dynamics of the system are issues that need to be faced to enhance flood prediction. In distributed models, the parameter values are physically

---

[1] This chapter is an edited version from, *"Alejandra María Rojas González, 2012. Flood prediction limitations in small watersheds with mountainous terrain and high rainfall variability. Unpublished PhD Thesis for Department of Civil Engineering and Surveying, University of Puerto Rico – Mayagüez Campus".*

[2] Numbers in brackets refer to the references at the end of this book.

based and the watershed is represented by grids, which approximates the parameter distribution and the initial conditions of the system. The modeler assigns the grid size resolution to the model, rainfall input scales and parameter values in a subjective way; subjective because the modeler has to select among various methods available for assigning grid point values (e.g., slope), and each method can influence the hydrologic result of the model. Each parameter and input are spatially and temporally scale-dependent, probability distributions are not known a priori, and the implications, in terms of uncertainty propagation though the system, are well understood.

This research study in Part I provides a guide for the modeler to develop a hydrologic model knowing the implications of scale and parameter uncertainties on the flow response in small watershed where the uncertainties affect more the prediction and answers several important research questions. An objective of this research was to address the following three research questions indicated in Chapter 1:

RQ1. How flood prediction is affected by the spatial variability of point rainfall at scales below that of the typical resolution of radar-based products?

RQ2. How does parameter resolution affect the model's predictive capabilities and the errors of the hydrologic system?

RQ3. Would the assumptions developed for the small scale enhance the hydrologic predictability at larger scales?

The main conclusions that can be drawn from this research are presented below:

- Rainfall variability was measured in a mountainous area of 4×4 km^2 (16 km^2) using a high-density rain gauge network. High spatial variability over short distances was measured. The standard deviation increased with increasing rainfall depth and the trend slope line (coefficient of variation) between average rainfall and standard deviation increased with increasing area of coverage (from 4.5 to 16 km^2), [RQ1].

- NOAA's MPE (Multi-sensor Precipitation Estimation) product was evaluated in an area of 16 km^2 using the rain gauge network at hourly and daily time steps. MPE overestimated rainfall at the hourly time step and underestimated at the daily time step. Non-significances were found in the hit rate between time steps. The probability of detection (POD) by the radar increased with the time step from 0.57 (hourly) to 0.833 (daily). False alarm rates were reduced with the larger time step, [RQ1].

- Large biases were found in the hourly time step and are associated with small rainfall detections and the resolution of both instruments. The bias between radar and the rain gauge network was event and time dependent. It is a random variable and follows a normal with box-crox distribution, [RQ1].
- Hydrologic predictability was studied as influenced by rainfall resolution inputs and hydrologic model resolutions, indicating their respective effects on flow response. The May 2 and September 8, 2008 events produced the greatest total average rainfalls and standard deviations, with high and low values of 5 days antecedent rainfall, respectively. No significant changes in total storm rainfall were observed with the interpolations at different scales, but produced important differences in rainfall intensity changes cell to cell though time, [RQ1].
- The slope map is an important input to the model. Decreases in the average slope will delay the time to peak and reduce peak flows. Up-scaling methods were tested to conserve the average slope and Method 2 was recommended to upscale a slope map in mountainous basins with high elevation variability over short distances, [RQ2].
- Río Rosario watershed was most sensitive to overland roughness with a Sr average of −13.7 followed by channel roughness with −7.4, overland hydraulic conductivity with −3.3 and initial soil moisture with 2.8 for peak flow. Sr for Río Grande de Añasco and Río Guanajibo watersheds indicate that the most sensitive parameters were channel roughness with −13.8 and −19.0, respectively, followed by overland roughness with −8.5 and −10.6 and initial soil moisture with 6.6 and 6.1, respectively, [RQ2].
- Río Rosario, Grande de Añasco and Guanajibo watersheds were most sensitive to initial soil moisture followed by overland hydraulic conductivity and soil depth for runoff depth, [RQ2].
- Variations between events can change the ranking of the input parameters studied. This was observed in the case of both variables (peak flow and runoff depth) indicating time or event dependence in Sr computations related to antecedent soil moisture, [RQ2].

- Rainfall ensembles for different resolutions were evaluated and a guide was presented in which the modeler can decide or to know the uncertainties associated with each resolution. In general, the rainfall ensemble at 100 m, followed by 400 m and 200 m can represent very well the peak flow, volume and time to peak, three variables that indicate a good agreement between the observed hydrograph and the prediction, [RQ1, RQ2, RQ3].
- Hydrographs that present various bumps during the event can be represented very well with the hydrologic model at 10 m grid size spacing, locating the time to peak with the corresponding peak flow. However, this grid size has problems with volume computations for dry conditions. Another hydrologic model that can capture the bumps is the 100 m grid size spacing and can produce the results for runoff depth very well, [RQ2, RQ3].
- Based on the analyzes presented in this research, the recommended up-scaled rainfall resolution, which will provide equivalent accuracy with the 100 m rainfall resolution, is 1000 m. The recommended up-scaled grid resolution, which will provide equivalent accuracy with the 10 m resolution, is 100 m, [RQ1, RQ2].
- Another useful result, but not specifically related to any of the research questions, pertains to the estimation of potential evapotranspiration (PET). The temperature/elevation linear regression equations of Goyal et al. [40] were evaluated to calculate the PET at a daily time step using the Hargreaves-Samani equation [45] and the results showed similar regression coefficients between observed and calculated T_{max}, T_{min} and T_{ave} values with the temperature/elevation lineal regression equations by Goyal [40]. The most sensitive parameter is the solar radiation, because the temperature model [40] cannot represent the spatial variability of this parameter using the daily interpolation for extraterrestrial radiation and the T_{max} and T_{min} calculated with the elevation model. Therefore, the use of Eq. (2) in Chapter 3 is recommended with measured values of solar radiation and temperature values either measured or estimated using the Goyal et al. [40] method.

For future works, it is recommended to include more events in the analysis for the TBSW, covering different event types, magnitudes and antecedent soil moisture condition as was covered in this research, from dry to wet conditions. Including more events would validate the findings in this research.

The Part I includes Bias as an additional perturbation parameter, using a normal with Box-crox transformation (lambda = 0.15) probability distribution function, to evaluate the uncertainty propagation though the hydrologic model.

The methodology used in this research to evaluate the rainfall resolution impact on hydrologic response using the bias corrected MPE product, could be reevaluated using the CASA radar data (when available) with high-resolution grid size to decide which resolution is desirable from a hydrologic point of view.

Currently, a high-density rain gauge network, extending over the MBDB area, which could be used to validate the NEXRAD rainfall estimates, does not exist. In the near future, it is hoped that this rainfall resolution gap will be filled by the CASA radars and that the hydrologic model formulated can be tested.

9.2 SUMMARY OF RESEARCH STUDY IN PART I

An evaluation of the interrelation between different up-scaling parameters and inputs were evaluated to quantify their influence on hydrologic predictability in complex terrain and small watersheds. An up-scaling experiment was performed, consisting of increasing the grid size to produce incrementally coarser resolution maps of each parameter, terrain and rainfall inputs. Each resolution was evaluated by an ensemble approach and generalized likelihood uncertainty estimation (GLUE) methodology using high-resolution rain gauge network (rainfall resolution of 100 m) and fully distributed hydrologic model (10 m). Each parameter perturbation, hydrologic model resolution, and rainfall resolution combination were modeled producing deterministic forecasts called "ensemble members".

Objective functions were used to evaluate the behavior of each ensemble with observed data using the variables time to peak, runoff depth and peak flow observations. Ensemble skill was evaluated using scalar measures of accuracy for continuous prediction as mean absolute errors (MAE), root mean square error (RMSE) and bias between the average ensembles to observation variable. Probabilistic distribution functions (PDF) were generated for each ensemble and prediction skill was measured by ranked probability score (RPS). Based on the analyzes presented in this research, the recommended up-scaled rainfall resolution, which will provide equivalent accuracy with the 100 m rainfall resolution, is 1000 m, and the recommended up-scaled hydrologic model grid resolution, which will provide equivalent accuracy with the 10 m resolution, is 100 m.

ACKNOWLEDGEMENTS

Author thanks the CASA project supported by the Engineering Research Centers Program of the National Science Foundation under NSF award number 0313747, for their financial support during the graduate studies, in the form of a graduate research assistantship, conferences, retreats and a summer internship to the National Weather Center at Oklahoma; the UPRM NOAA-CREST project (grant NA06OAR4810162) for financial support and NOAA-CREST students for collecting and maintaining the rain gauge data, which was used in this project. Special thanks to Dr. Eric Harmsen for his continuous support, the opportunity to research under his guidance and supervision, for the motivation, encouragement, reviews and his wonderful friendship. Thanks to Dr. Baxter and Jean Vieux for their support in the Vflo model; Dr. Sandra Cruz Pol and Dr. José Colom for the opportunity to participate and perform the research on CASA project.

KEYWORDS IN PART I

- **actual vapor pressure**
- **adjustment factor**
- **air temperature**
- **Annual Maximum Series**
- **ANOVA**
- **bias**
- **CASA project**
- **CASA Student Testbed area**
- **CASCade 2 Dimensional Sediment**
- **Centro de Recaudación de Impuestos Municipales**

- channel bed slope
- channel velocity
- coefficient of determination
- Collaborative Adaptive Sensing of the Atmosphere
- conceptual rainfall-runoff
- contingency table
- contributing drainage area
- Cooperative Remote Sensing Science and Technology Center
- cumulative distribution function
- cumulative forecast
- cumulative probability
- data file in text format
- decibels
- Digital Elevation Map
- discrete bias
- Distributed Hydrology Soils and Vegetation Model
- ensembles
- evapotranspiration
- evapotranspiration crop coefficient
- exponential weighted
- extraterrestrial radiation
- false alarm
- false alarm rate
- Federal Emergency Management Agency
- finite element method
- flood alarm system
- flood insurance study
- flow area
- flow depth
- forecast
- Gaussian distribution function

- generalized likelihood uncertainty estimation
- generalized Pareto distribution
- geographical information systems
- GLUE
- hit rate
- hydro estimator
- Hydrologic Engineering Center
- Hydrologic Engineering Center – River Analysis System
- Hydrologic Modeling System
- Hydrologic Rainfall Analysis Project
- inverse distance weighting
- kinematic wave analogy
- land sat
- Manning's roughness factor
- maximum air temperature
- Mayagüez Bay Drainage Basin
- mean
- mean absolute error
- mean annual rainfall
- minimum air temperature
- multisensor precipitation estimation
- National Oceanic and Atmospheric Administration
- net radiation
- next generation radar
- normal distribution function
- North American Datum
- ordered physics-based parameter adjustment
- partial duration series
- peak flow
- Pearson correlation
- Penman Monteith equation

- physically based distributed
- physically based hydrologic developed by Vieux and Associates, Inc.
- potential evapotranspiration
- precipitation radar product
- probability
- probability distribution function
- probability of detection
- probability plot
- psychometric constant
- Puerto Rico
- Puerto Rico Water Resources and Environmental Research Institute
- Puerto Rico Water Resources Authority
- quantitative precipitation estimation
- quantitative precipitation estimation and segregation using multiple sensors
- quantitative precipitation forecast
- radar rain rate equation
- rainfall
- rainfall rate
- ranked probability score
- reference evapotranspiration
- reflectivity
- relative sensitivity coefficient
- root mean squared error
- runoff
- saturated hydraulic conductivity
- saturated vapor pressure
- simulation
- slope of the vapor pressure curve
- Soil Climate Analysis Network

- soil heat flux density
- Soil Survey Geographic Database
- solar radiation
- standard deviation
- stream volumetric discharge
- subwatershed
- Systeme Hydrologique European
- Testbed Subwatershed
- Thematic Mapper
- time to peak
- Tropical Agriculture Research Station
- Tropical Rainfall Measuring Mission
- U.S. Army Corps of Engineers
- U.S. Geological Survey
- United Stated Department of Agriculture
- upscaled
- vapor pressure
- variance
- Vflo
- wind velocity

PART II

FLOOD ALERT SYSTEM USING HIGH-RESOLUTION RADAR RAINFALL DATA

CHAPTER 10

FLOOD ALERT SYSTEM USING HIGH-RESOLUTION RADAR RAINFALL DATA: INTRODUCTION[1,2]

LUZ E. TORRES MOLINA

CONTENTS

10.1   INTRODUCTION

Chapters 10 to 20 in Part II of this book volume will present in detail the research study on flood alert system using high-resolution radar rainfall data.

Portions of western Puerto Rico are subject to flash flooding due to sudden, extreme rainfall events, some of which fail to be detected by NEXRAD radar located approximately 120 km away in the town of Cayey, Puerto Rico and partially obstructed by topographic features. The use of new radars with higher spatial resolution and covering areas

---

[1] This chapter is an edited version from: *"Luz E. Torres Molina, 2014. Flood Alert System Using Rainfall Data in the Mayagüez Bay Drainage Basin, Western Puerto Rico. PhD Thesis, Department of Civil Engineering and Surveying, University of Puerto Rico, Mayagüez Campus".*

[2] Numbers in brackets refer to the references at the end of this book.

missed by the NEXRAD radar, are important for flood forecasting efforts, and for studying and predicting atmospheric phenomena.

Recently, Trabal et al. [85] at the University of Puerto Rico – Mayagüez Campus initiated investigations using two types of radars, namely: Off-the Grid (OTG) and TropiNet, with radius of coverage of 15 km and 40 km, respectively. This network will monitor the lower atmosphere where the principal atmospheric phenomena occur. This study indicates for the first time that TropiNet radar technology can be used for hydrologic analyzes and specifically for rainfall forecasting in western Puerto Rico.

Short-term rainfall forecasts have commonly been made using *Quantitative Precipitation Forecasting* (QPF). The introduction of QPF in flood warning systems has been recognized to play a fundamental role. QPF is not an easy task, with rainfall being one of the most difficult elements of the hydrological cycle to forecast [24], and great uncertainties still affect the performances of stochastic and deterministic rainfall prediction models [86]. Currently, this capability does not exist in western Puerto Rico, and it is needed because of the potential for flooding in certain areas (e.g., in flood plains near the principal rivers of the region).

The Part II on flood alert system using high-resolution radar rainfall data includes a research study where short-term rainfall forecast analysis is performed using nonlinear stochastic methods. Once obtained, the rainfall forecast is introduced into a hydrologic/inundation model *Vflo* and into the Inundation Animator configured for the Mayagüez Bay Drainage Basin (MBDB). Specific components of the research in this Part II are:

- The inclusion of calibration and validation of rainfall estimates produced by the TropiNet radar network,
- The development and validation of the stochastic rainfall prediction methodology,
- The calibration and validation of the inundation algorithm at selected locations within the MBDB, and
- The proto-type of an operational, real-time flood alarm system for the MBDB. The proto-type, automated Flood Alarm System (FAS) will be able to send near-real time updated inundation images to a website on the Internet.

10.2 OBJECTIVES

The prediction or forecast of natural disasters is critically important for emergency management workers. An aim of forecasting is to gain adequate time to evacuate people from disaster zones, to minimize loss of life, and reduce damage to structures and infrastructure, and minimize economic loss. Every country around the world is exposed to various types of natural disasters depending on its geographical location (e.g., tornados, hurricanes, volcanoes, earthquakes, flash floods, hail, snow, drought, tsunamis, fire and others).

Flash flooding is defined as the rapid rise in water level causing flooding of an area. It may be caused by heavy rainfall associated with storms, hurricanes or tropical storms. The World Meteorological Organization (WMO) has defined the flash flood as "*a flood that follows the causative precipitation event within 6 h of time* [113]". The National Weather Service has estimated that more than 70% of flash flood warnings may be issued with less than a one hour lead-time and that more than 50% of flash flood occurrences allow no lead-time whatsoever (personal communication, Ernesto Rodriguez, NWS, San Juan [57]).

Small watersheds have short time of concentration and reaching its peak, the response time of smaller basins could be on the order of a few hours or less than one hour. Recently in Puerto Rico, flash flooding has occurred from some significant rainfall, events that can occur over very short time scales (e.g., one to several hours on September 22, 2008, over almost whole of Island). The susceptibility of the island of Puerto Rico to flooding is high due to a variety of factors including its mountainous runoff that drains into flat floodplain terrain with poor drainage, intense rainfall and urban development, and the variability of rainfall in the island is a huge argument to use radars in the precipitation forecast [56].

Lately, some researchers have been using radars and they have indicated that a key factor for accurate flood estimates and forecast is accurate rainfall for input to the hydrological model. Rainfall data are traditionally obtained from an often-sparse network of rain gauges that may not record the rainfall event with adequate spatial and temporal scales. Especially during heavy convective storm, a significant rainfall occurs over a limited areal extent and the weather radar has enormous potential in this field, with high spatial resolution and temporal continuity [83].

The rainfall forecast is an important component of the flood alert system that is designed to collect, handle, analyze and distribute information for the purpose of providing advanced warning of a flood condition. This is possible when there is a good stochastic rainfall prediction, and a good hydrological model. The most important hydrologic model outputs are: the predicted peak flow, runoff volume and time to peak. These parameters are dependent on the quality of the hydrologic model and the rainfall estimated from the rain gauge or radar network [56].

To manage these conditions, the current study seeks the use of the high-resolution rainfall from the TropiNet radar network that will provide an excellent source of rainfall data, previously not available for short-term rainfall and flood prediction studies in Puerto Rico. Furthermore, the current study introduces the application of a novel nowcasting model, improvements in the accuracy of short-term rainfall forecast due the high spatial and temporal resolution in radar rainfall data, technology for real-time inundation mapping, which has not been used in previous flood prediction studies in Puerto Rico, and which will be a powerful new tool in the hands of emergency flood management personnel. The methodology proposed in this research can be applied to other watersheds in Puerto Rico or in others regions within the tropics.

Rainfall forecast and their integration into the disaster plan can have social and economic benefits, with a lead-time adequate to allow evacuation from flood prone areas within the Mayagüez Bay Drainage Basin (MBDB).

Therefore, the interest of this study was to develop a forecasting model for the prediction of short-term rainfall in the western Puerto Rico. The forecast results were then introduced within a hydrologic model *Vflo* and an Inundation model "*Vflo* Inundation Animation" to get the animation of the flow into the rivers. Specific study objectives were:

a. Analyze the rainfall structure and behavior to develop an accurate stochastic model to forecast short-term rainfall for selected areas within the MBDB. The results using the forecasting model were compared and analyzed statistically with the observed data for validation and calibration of the model.

b. Apply the forecast rainfall data to the hydrologic model Vflo and Inundation Analyst module to obtain animation of flooding at selected locations within the MBDB in real time and to compare the results of the rainfall and hydrologic forecasts with observed data.

c. Develop a proto-type real-time flood forecast alarm system for the MBDB.

CHAPTER 11

FLOOD ALERT SYSTEM USING HIGH-RESOLUTION RADAR RAINFALL DATA: A REVIEW[1, 2]

LUZ E. TORRES MOLINA

CONTENTS

11.1   STOCHASTIC MODELING AND SHORT-TERM RAINFALL FORECASTING

This chapter presents the literature review on flood alert system using high-resolution radar rainfall data. There are many approaches that can be used to predict the future direction and magnitude of a physical process, such as rainfall. Forecasting is a large and varied field having

---

[1] This chapter is an edited version from: *"Luz E. Torres Molina, 2014. Flood Alert System Using Rainfall Data in the Mayagüez Bay Drainage Basin, Western Puerto Rico. PhD Thesis, Department of Civil Engineering and Surveying, University of Puerto Rico, Mayagüez Campus"*.

[2] Numbers in brackets refer to the references at the end of this book.

two predominant branches: Qualitative Forecasting and Quantitative Forecasting [38]. Quantitative Forecasting should satisfy two conditions, the accessible numerical information about the past and assumptions that some aspects of the past patterns will continue into the future. Quantitative Forecasting can be divided into two classes: time series and explanatory models. Explanatory models assume that the variables to be forecasted exhibit an explanatory relationship with one or more other variables, in contrast, time series forecasting uses only information on the variable to be forecasted, and make no attempt to discover the factors affecting its behavior [38]. The time series models attempt to capture past trends and extrapolate them into the future. There are many different time series models but the basic procedure is the same for all as illustrated in Figure 11.1.

Some of the most common time series methods include: Autoregressive, Moving Average, Exponential Smoothing, Autoregressive Moving Average, Extrapolation, Linear Prediction and others [5]. This research includes a new type of time series nonlinear with stochastic and deterministic components, which will be explained later.

The autoregressive (AR) method is a type of random process, which is used to predict some types of natural phenomena, falling within the group of linear prediction formulas. The moving average (MA) method is a way where the current observations depend on all past observations [5]. Exponential smoothing is a popular scheme to produce a smoothed time series; exponential smoothing assigns exponentially decreasing weights to the observations as they get older. That is to say recent observations are given relatively more weight in forecasting than the older observations.

With the Autoregressive Moving Average (ARMA) method, models are used to describe stationary time series, which represent the combination of an autoregressive (AR) model and moving average (MA) model. The order of the ARMA model in discrete time (t) is described by two integers (p, q), which are the orders of the AR and MA parts, respectively.

FIGURE 11.1   Flowchart for a stochastic model [5].

A process is considered to be stationary when parameters, such as the mean and variance, do not change over time or maintain the same range. Autoregressive (AR) or ARMA are the models that are widely used in the prediction of flows in water resources. Other time series methods include extrapolation and linear prediction, and the nonlinear prediction with exponential component, depending of data behavior.

A time series is part of a stochastic process. The word stochastic comes from the Greek *stokhastikos*, an adjective that refers to system whose behavior is intrinsically nondeterministic, or sporadic, and categorically not intermittent [48]. The stochastic methods are techniques used in prediction of events, such as: winds, hurricane tracks, temperature, humidity, rainfall, floods, etc. The stochastic concept has been used in hydrology since the beginning of twentieth century [74] applied in the river flow sequence analysis, but only in the 1970's were autoregressive models applied to seasonal and annual hydrologic time series. Research in hydrologic time series has aimed towards studying the main statistical characteristics, providing physical justification for some stochastic models, developing new models, improving existing modeling parameters, developing new modeling procedures, improving tests of goodness of fit and other parameters applied to hydrology [74].

Forecasting is a relatively new science within hydrology and the atmospheric sciences [75]. Its application has led to the reduction in deaths caused by natural disasters. "The Time Series Analysis" of Box and Jenkins [5] constitutes an important contribution to the field of stochastic analysis for the purpose of forecasting hydrologic phenomena. The book focuses on the application of the autoregressive and moving average models for forecasting.

A number of researchers have developed hurricane prediction tracking models in Puerto Rico. For example, Ramirez-Beltran [62] used historical data to develop a stochastic model to predict the behavior of hurricane tracks. The parameter estimation scheme, based on recursive and iterative algorithms, used historical records for hurricanes to fit vector autoregressive models. The identified models have been classified according to the order of the model. The first observations of a given hurricane are compared with historical hurricane tracks. Ramirez-Beltran [62] concluded that the vector ARMA model has an excellent potential and may help reduce official forecasting error compared with a Statistical-Dynamical Hurricane

Track Prediction Model (NHC90) from the National Meteorological Center. Ramirez-Beltran et al. [64] introduced a rainfall forecast methodology based on NEXRAD data that was used as the basis to formulate the new rainfall methodology.

The ideal forecast rainfall is based on the meteorological analysis but this is not always available, when this information is not accessible, the forecast rainfall can be based on current and past rainfall. The forecasting of rainfall has been investigated by Burlando et al. [8]. Their research relates on whom to forecasting rainfall at a point, with a simple formulation. Various models can be used for representing forecasting at point. Several models have been developed which describe storm arrivals following a Poisson process. However, the formulation for real time rainfall forecasting based on these models is too complex [65].

Burlando et al. [8] discussed forecasting of short-term rainfall using ARMA models defined at 1 h and 2 h of time scale. They suggested that parameter estimation models based on short-term precipitation records defined at hourly time-scale is more complex than when data is defined at longer time periods such as months. They forecasted rainfall by assuming that hourly rainfall follows an ARMA process. This assumption is based on the fact that the autocovariance structure of some point processes, such as the hourly rainfall.

Burlando et al. [8] investigated two estimation and fitting procedures. The first takes all rainfall occurrences throughout the period of record as the basis for parameter estimation, thus a given set of parameters results are obtained for a given month or season, and the second is an event-based estimation approach, each storm or independent rainfall of the month or season is considered separately for parameter estimation. Thus a different parameter set was determined for each storm or rainfall event considered. These procedures were compared for rainfall data at a point and rainfall data averaged over the basin. The analysis used hourly rainfall from two gauging stations in Colorado, USA and from some stations in Central Italy. Their research is related to forecasting rainfall at a point using rain gauges, Thyessen polygons were used to weigh contribution from each rain gauge. The results show that the event-based estimation approaches yields better forecasts than the continuous approach and is capable of producing the rainfall intensity fluctuations (Figure 11.2).

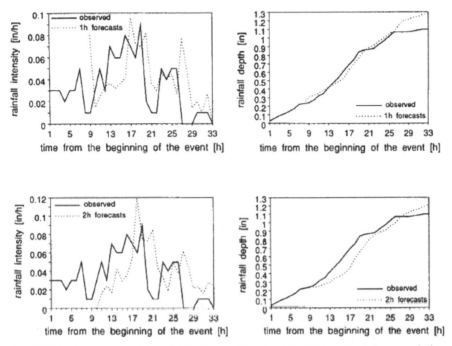

FIGURE 11.2   Forecast 1 h and 2 h ahead of hourly rainfall intensity and accumulative rainfall. Using event-based approaches for the event of 18 February 1953, Denver station, Colorado, USA [8].

Burlando et al. [8] assumed that the rainfall processes are typically nonstationary and skewed. To circumvent nonstationary, the rainfall data are grouped by month of season; thus the model is applied separately for data of a given month or a given season. Accordingly, model parameters such as autoregressive and moving average coefficients were determined from precipitation data pertaining to a given month or season only. Parameter estimation of ARMA models based on short-term precipitation records defined at hourly time-scales is more complex than when data is defined at longer time periods such as months. The main reason is the intermittent characteristic of hourly precipitation. Therefore, two alternative approaches were followed by Burlando et al. [8] for parameter estimation and forecasting. The first was referred to as the continuous approach and the second as the event based approach.

In the continuous approach, all the precipitation events which occurred in a given month or season were considered for parameter estimation.

Thus, a given set of parameters results for a given month or season. In the continuous approach, the precipitation data used for parameter estimation were arranged in two different ways. In the first, no differentiation was made between storm events, and the whole dataset, including zero recorded precipitation, was used for estimation.

In the event-based estimation approach, each storm event regardless of the month or season is considered separately for parameter estimation. Thus, a different parameter set was determined for each storm event considered. On the other hand, in the event-based approach, as only data of the current storm is used for estimation, the number of observations is small, and especially at the beginning of the storm event when only a few rainfall measurements are available. The influence of the storm movement was not considered here, and the scale of temporal and spatial aggregation at which data should be monitored are two factors which they may improve the reliability of rainfall forecasts [8].

Delleur and Kavvas [15] used the autoregressive moving average (ARMA) model to study the average rainfall time series over 14 basins. Results showed that the model is adequate for a short-range forecast at one or two time steps ahead. They claimed that due to the rotation of the earth around the sun, the monthly rainfall time series exhibit a yearly periodicity. The time series models are usually fitted to the stationary random component of the spectrum or equivalently to the decaying part of the autocorrelation function. They said that this is necessary to remove the nonstationary component of the process.

Delleur et al. [16] used a model that included a Markov chain to simulate the sequences of dry and wet days. They found that the models simulated the sequences of dry and wet days well. However, the amount of daily rainfall was not described adequately.

McLeod [47] demonstrated that the principle of Parsimony is helpful in selecting the best model for forecasting river flow. His work demonstrated the importance of model adequacy for seasonal river flow and incorporated seasonal periodic correlation. Briefly, their experience with river flow time series suggests that the best forecasting results are obtained by following the general model building philosophy implicit in Box and Jenkins [5] with suitable modifications and improvements. Box and Jenkins [5] gave a method to estimate the orders of the AR and MA terms of a model based on autocorrelations and partial autocorrelations.

A popular decision rule for comparing models in the time series litera-ture is the Akaike Information Criterion (AIC) [1]. This criterion is known as the test for the Parsimony of parameters. Several investigations have used AIC criterion for choosing the model type, order and in constructing an appropriate model for a given streamflow series. The following proce-dure is usually followed:

a. The appropriate type of model, among AR, MA, ARMA, ARIMA (autoregressive integrated moving average) and seasonal ARIMA models is selected.
b. The choice of order for the selected model is determined.
c. The parameters in the model using the given stream flow series are estimated; and validation of the model by residual testing and by simulation is performed. This procedure is applied to identify models for forecasting and synthetic generation.

Mujumdar and Nagesh [50] used two criteria for the model selection, Maximum Likelihood rule (ML) and Mean Square Error (MSE) are used for the selection of the best model for each of the rivers considered. The selec-tion of a model by the ML rule involves evaluating a likelihood value for each of the candidate models and choosing the model which gives the highest value. In general, as the number of parameters increase in the function, the likelihood value decreases. Thus it is to be expected that the ML rule selects models with a small number of parameters, this is the principle of parsimony propounded by Box and Jenkins [5]. The maximum likelihood estimation cri-terion is suited for the selection of a model for simulation purpose. For short-term forecasting, such as one step ahead forecasting, the mean square error (MSE) criterion may be more useful [40]. Selection of a model based on an MSE criterion is quite simple and can be summarized as follows:

a. Estimate the parameters of different models using a portion, usually half of the available data.
b. Forecast the second half of the series one step ahead by using the candidate models.
c. Estimate the MSE corresponding to each model and
d. Select the model that results in the least value of the MSE.

For all cases presented by Mujumdar and Nagesh [50], the simple model AR resulted in the minimum value of the MSE, underlining the fact that for one step ahead forecasting, quite often the simplest model is sufficient. Additionally, the case study revealed that as the number of parameters increased, the MSE increased, which is an interesting result contrary to the common belief that models with larger number of parameters, give better forecasts. For all series of the stream flows considered, the AR model is strongly recommended for use in forecasting the series one step ahead. Salas and Obeysekera [75] worked with time series models in streamflow and have stated that data generation and forecasting of seasonal streamflow are often needed in the planning and management of water resources systems. Both data generation and forecasting are based either on stochastic models alone or in combination with corresponding conceptual models of the system under consideration. In most cases, stochastic model are usually developed, based on the available data on hand.

In modeling time series of annual flows, the assumption of stationarity of the series is usually made, so that stationary stochastic models can be applied in the study. When dealing with time series of seasonal flows, the modeling is more complex. The main reason is the inherent periodicity in several statistical characteristics that invariably lead to stochastic models with periodic parameters. Most techniques available for diagnostic checks have been limited to stationary models, although some approximations have been suggested for models with periodic parameters. A number of conceptual simulation models representing the hydrologic cycle of watersheds have been suggested in the literature since early 1960 s. Examples of such models are the new version of the Stanford watershed model [39] and the Sacramento model of the National Weather Service [9]. Extensive literature already exists on these two modeling approaches. However, less attention has been given to linking both conceptual and stochastic modeling schemes.

Kohnova et al. [42] conducted a study involving the modeling and forecasting of discharge and rainfall time series in the area of the Klastorske Luky wetland – Slovakia. They first analyzed the systematic components (trends, seasonality, periodicity and residual components). Subsequently, prediction models for the mean monthly discharges and the mean monthly precipitation totals were derived. The models tested were linear ARMA models. The results obtained could help ecologists in making decisions on wetland management, improving the ecological conditions in the analyzed wetland, and planning future ecotechnical measures.

Many problems related to water resources and environmental systems deal with temporal data which need to be analyzed by means of a time series analysis, which has become a major tool in hydrology. Time series analysis is used for building mathematical models to describe hydrological data, forecast hydrologic events, detect trends, provide missing data, etc.

After analyzing some types of autoregressive models, Kohnova et al. [42] concluded that the ARMA model can be used to generate synthetic traces of monthly rainfalls, particularly useful in the analysis of water resources projects on basins.

Katz and Skaggs [41] studied statistical problems that may be encountered in fitting ARMA processes to meteorological time series. They used techniques that lead to an increased likelihood of choosing the most appropriate ARMA process to model the available data are emphasized. ARMA models are well suited to the analysis and forecasting of time series which are by nature or by manipulation persistent and thus, are especially useful in climatological analysis.

Box and Jenkins [5] are primarily responsible for making readily accessible the necessary statistical methodology for applying ARMA models to real data and for taking advantage of the use of these models in forecasting. While ARMA processes have many advantages over other somewhat similar processes, their application to modeling meteorological data may require an increased degree of mathematical sophistication on the part of the researcher.

Other examples of rainfall forecasting models were developed and available in the literature. Prediction of Rainfall Amount Inside Storm Events (PRAISE) is a stochastic model developed by Sirangelo et al. [80] to forecast rainfall height at site. PRAISE is based on the assumption that the rainfall height accumulated on a delta time is correlated with a variable that represent antecedent precipitation. The mathematical background is given by a joined probability density function and by a bivariate probability distribution, which is referred to the random variable that represents rainfall in a generic site and antecedent precipitation in the same site. The peculiarity of PRAISE is the availability of the probabilistic distribution of rainfall heights for the forecasting hours, conditioned by the values of observed precipitation. The Calabria region in southern Italy was selected to test performances of the PRAISE model [80]. PRAISE was applied to all of the telemetering rain gauges of the Calabria region, in Southern Italy. The calibration model showed that the hourly rainfall series present a constant value of memory equal to 8 h, for every rain gauge of the Calabria network.

11.2   NOWCASTING

The interest from rainfall forecast with high spatial and temporal resolution has been increased in contemporary days. Only some equipment like radars is capable of producing high spatial and temporal resolution. Early algorithms were based on pattern recognition of rainfall echoes from which cross-correlation coefficients can be calculated and used to predict the motion of the storm feature [21].

Dixon and Wiener [20] developed a nowcasting system called TITAN (Thunderstorm Identification Tracking Analysis Nowcasting) to predict convective rainfall. TITAN uses real-time automated identification tracking and short-term forecasting of storm which besides is able to nowcast storm development and movement.

Nowcasting can be described as the production of short-term (0–3) hours lead-time precipitation forecasts based mainly on the extrapolation of future data from current radar data images [81]. Nowcasting has benefited many different fields in addition to flood forecasting, including more general public weather warnings, water management, storm sewer operation, and irrigation, wet deposition of pollutants, construction site management, and transportation systems [7].

11.3   RADAR RAINFALL ESTIMATION AND VALIDATION

The National Weather Service is incharge of providing weather, hydrology, and climate forecasts and warnings for the United States including Puerto Rico and U.S Virgin islands, working with a network of 159 high-resolutions Doppler weather radars, commonly referred to as NEXRAD (Next-Generation Radar). The technical name for NEXRAD is WSR-88D, which stands for Weather Surveillance Radar, 1988, Doppler [52]. NEXRAD detects precipitation and atmospheric movement or wind. The NEXRAD radars can provide information that can help mitigate disasters caused by flash floods. Errors can occur with the methodology for observations far from the radar, where the earth's curvature limits the observation of the lower atmosphere (Figure 11.3).

NEXRAD coverage has limitations in observing below 10,000 feet or 3 kilometers (called the Gap) above sea level for the Mayagüez area and nearby towns [12]. At these locations, NEXRAD cannot "see" if raindrops are forming within the Gap, resulting in a different rain rate than other radars which can measure the lower portion of the cloud (OTG and TropiNet). In the OTG and TropiNet radars, the rain rate equations can be selected, whereas NEXRAD rain rate uses the tropical equation with a threshold reflectivity (Z) of 53 dBZ, Z values above 53 dBZ are assumed to be hail and are not considered [73]. Other difference between NEXRAD and TropiNet radar is that NEXRAD has Doppler capabilities given information on cloud motion, and TropiNet has Polarimetric capabilities which give information on precipitation type and rate. Polarimetric radars refer to dual-polarization radars which transmit waves that have horizontal and vertical orientations. The horizontal wave provides a measure of horizontal

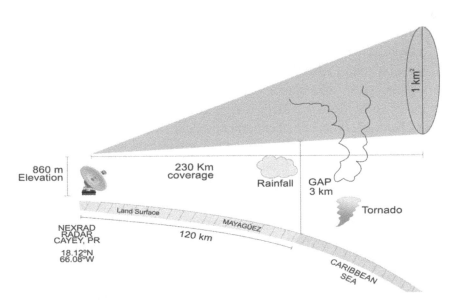

FIGURE 11.3 Long-range problems with NEXRAD, based on *Westrick* et al. [111]. Note: The figure does not include topography of the land surface.

dimension of the cloud and rainfall where the vertical wave provides a measure of particle size, shape and density.

The use of the new radars OTG and TropiNet with higher spatial resolution and their observations of the lower atmosphere in the western Puerto Rico area can provide better atmospheric information in the lower zone because curvature effect is minimal, at minimum elevation [12]. The OTGs radars have been developed based on the modification of off-the-shelf marine radars, which are characterized by low power consumption (~180 Watts), short range (15 km) and low cost (~$30,000) [12]. The OTG radars are capable of operating independently of the existing power grid and communication infrastructure.

On January 2010, the OTG Radar No. 1 was successfully installed at the PR-1 radar tower on the rooftop of the Stefani Engineering Building at the University of Puerto Rico, Mayagüez campus. The radar has an estimated sensitivity of 12 dBZ at 15 km, a range and a mean cross-beam resolution of 120 m and 500 m, respectively, and is a 4 kw X-band marine radar [12]. This technology was developed by the Student Test Bed of the NSF Engineering Research Center for Collaborative Adaptive Sensing of the Atmosphere (CASA) in Mayagüez, Puerto Rico. *Arocho* et al. [3] conducted a preliminary calibration of estimated rain rates on the OTG Radar No. 1.

Recently, new radars (TropiNet-1) were installed in Cornelia hill (Guanajibo) and (TropiNet-2) in Lajas, while another will be installed in Isabela (UPR- agricultural Exp. Station). Previous known project as Puerto Rico Student Test Bed, is now part of the Puerto Rico Weather Radar Network (http://weather.uprm.edu). The RXM-25 radar is referred to as TropiNet because of the name of the project, and is designed to cover a range between 30 and 50 km at very high sampling resolution spatial 60×60 meters and temporal one minute that offers state-of-the-art radar data products. The RXM-25 is prepared to operate as a single radar unit or as a radar network, allowing both manual and automated control while the radar allows a motion over the whole hemisphere. Additionally, it uses a low operating cost magnetron transmitter capable of delivering up to 12 watts of average power per polarization channel. The RXM-25 is designed for easy access and maintenance, all of its signal processing and radar control software runs on a single server. Due to these characteristics, it is possible that the RXM-25 will provide the best overall data in western Puerto Rico area to forecast important rainfall events [25].

11.4 HYDROLOGIC AND INUNDATION (FLOOD) MODELING

Numerical hydrologic models are commonly used to predict surface runoff from watersheds, estimate peak stream flow and stage elevation. These models fall within three main categories: lumped, semilumped and distributed models. The lumped model bulks all of the rainfall/runoff processes into a few watershed scale parameters. An example of this type of model is the Sacramento Soil Water Accounting System [9]. The advantage of the lumped type model is that they are relatively easy to configure and to use. The semilumped model allows for the distribution of parameters in a watershed within homogeneous hydrologic response units (HRUs). An example of a semilumped hydrologic model is the Precipitation Runoff Modeling System (PRMS) developed by the U.S. Geological Survey (USGS) [44].

The third type of hydrologic model is the numerically distributed model. The most common numerical methods used for this type of model are the finite difference or finite element methods. An example of a numerically distributed model and the one that is used in this research is *Vflo*, developed by Vieux [96]. Some hydrologic studies in Puerto Rico have used the *Vflo* model, including Vieux and Vieux [100] and Rojas [56].

Vflo uses radar rainfall data as hydrological input to simulate distributed runoff and is based on Geographic Information Systems (GIS) data. It provides high-resolution, physics-based distributed hydrologic modeling for managing water from catchment to river basin scale, the prediction of flow rate and stage can be made in every grid cell in a catchment, river or region, and the output is integrated with the *Vflo* -Inundation-Analyst module. This module along with the Digital Elevation Model (DEM) data can be used to show the extent of flooding superimposed onto a land map.

Rojas [69] used Vflo to evaluate the influence of the interrelation between different up-scaling parameters and inputs on hydrologic predictability for use in flood prediction in the Mayagüez Bay Drainage Basin. Based on the analysis, the recommended upscaled rainfall resolution, which will provide equivalent accuracy with the 100 m rainfall resolution, is 1000 m, and the recommended upscaled hydrologic model grid resolution is 200 meters.

Much of the data used by Rojas [69] for the MBDB was originally developed by Prieto [61] as part of a preliminary hydrologic regional conceptual model for the MBDB and implemented in an integrated, fully distributed, physically based, numerical model *Mike She* [17]. The fully integrated model was capable to simulate surface and groundwater flow within the MBDB.

11.5   REAL-TIME FLOOD FORECAST SYSTEMS

The USGS has developed the Real Time Flood Alert System (RTFAS) for Puerto Rico [94]. RTFAS is a web-based computer program, developed as a data integration tool, and designed to assist emergency managers to predict flooding of streams in Puerto Rico. RTFAS is available online at "Real Time Flood Alert System – http://rtfas.er.usgs.gov/". It should be noted that the system is limited to providing stage elevation data at the locations of the USGS stream gauges.

The National Weather Service (NWS) establishes Flash Flood Guidance estimates in real time based on the Sacramento soil moisture accounting model [9, 27]. The analysis allows for the development of curves that relate threshold runoff to flash flooding. Unfortunately, the model has not been successfully implemented in all of the island's watersheds. For example, the model is incapable of producing accurate results in some of the watersheds of south-eastern Puerto Rico (personal communication, Ernesto Rodriguez, NWS, San Juan [57]), perhaps owning to the fact that some streams in this area loose significant amount of their flow to the underlying superficial aquifers [18].

Sepúlveda et al. [79] developed a hydrologic model to forecast real-time rainfall runoff within the Carraízo reservoir basin. The model estimated water volumes at the reservoir from the rainfall and discharge data obtained from the network stations within the basin.

FLOOD ALERT SYSTEM USING HIGH-RESOLUTION RADAR RAINFALL DATA: METHODOLOGY[1, 2]

LUZ E. TORRES MOLINA

CONTENTS

---

[1] This chapter is an edited version from: "*Luz E. Torres Molina, 2014. Flood Alert System Using Rainfall Data in the Mayagüez Bay Drainage Basin, Western Puerto Rico. PhD Thesis, Department of Civil Engineering and Surveying, University of Puerto Rico, Mayagüez Campus*".

[2] Numbers in brackets refer to the references at the end of this book.

12.1   INTRODUCTION

This chapter discusses the methods that were used to study the flood alert system using high-resolution radar rainfall data.

The University of Puerto Rico at Mayagüez has a research weather radar network and a rain gauge network developed by Luz Estella Torres-Molina for research work. The radar network provides information with higher spatial and temporal precision. TropiNet has a 60 x 60 meter spatial resolution at every pixel and temporal resolution of 1 min. A flood warning model must be operated based only on the data available at the time of forecast. Only the radar can display data in real time. This is not possible using rain gauges but the rain gauges are used for data validation. Rain gauges based systems must have a dependable and redundant telemetry system that will accurately and efficiently transmit data a central location for processing. The Data from TropiNet radar is used for rainfall prediction in MBDB, using stochastic methods. Once the rainfall forecast is obtained, the use of hydrologic models is necessary for analysis of flooding in this area.

This project is the first attempt to implement new technology using high-resolution radars for performance of flood alert/warning systems. This research is focused at the western Puerto Rico and can be applied in general to other areas or regions with the same rainfall type with the corresponding hydrologic soil and coverage data.

12.2   STUDY AREA

The study area, which encompasses the MBDB, is 819.1 km² in area [56, 61] and is located in western Puerto Rico. The region has three important watersheds: Río Grande de Añasco, Río Guanajibo and Río Yagüez. The area includes 12 municipalities: Mayagüez, Añasco, Las Marías, San Sebastián, Lares, Maricao, Yauco, Adjuntas, Sabana Grande, San Germán, Hormigueros and part of Cabo Rojo. These three important rivers discharge into Mayagüez, Añasco and Cabo Rojo branches, respectively. According the U.S. Census Bureau, Mayagüez has 89,080 habitants and a total area of approximately 143.53 km² of which about 25.20 km² are in flooding areas; Añasco has 29,261 habitants with a total area of about 102.82 km² and 23.11 km² are in flooding areas; and Cabo Rojo has 50,917 habitants with a total area of about 187.81 km² and 44.42 km² are in flooding area [88], as shown in Figure 12.1.

The Río Grande de Añasco originates at the Cordillera Central, flows west and discharges into the Bahia de Mayagüez. The alluvial valley covers an area of approximately 46.62 km². It is bounded by hills to the north, east and south and by the Bahia de Añasco to the west. The major tributaries of the Añasco River that flow into the lower valley are the Rio Dagüey and the Rio Cañas. The basin is located in west-central Puerto Rico, in the municipalities of Añasco, Lares, Las Marías, Maricao, Mayagüez and San Sebastián.

12.2.1 BASIN OF THE RÍO GRANDE DE AÑASCO

The basin of the Río Grande de Añasco has an area of 467.7 km² of which approximately 10% of the area is flat land and the other remaining 90%

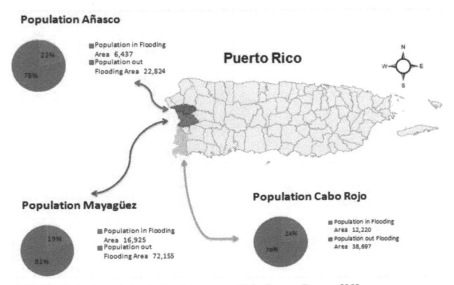

FIGURE 12.1 Population in floodable areas, U.S. Census Bureau [88].

is mountainous. The floodplain covers approximately three-fourths of the flat land. The residential developments in the Añasco municipality are partially within this area, and therefore can be affected by flooding. Río Grande de Añasco flows westerly 74 km to the coast, where its discharges into the Bay of Mayagüez. Changes in elevation (DEM model) are shown in Figure 12.2 and vary from zero meters at mean sea level in the coastal areas to 960 meters in the mountainous areas. The upper reaches of the basin contain four interconnected reservoirs: the Lago Toro, Lago Prieto, Lago Guayo and Lago Yahuecas, to the Añasco watershed downstream of the lakes which is not significant for regional water budget estimation [61]. These transport outside water to Lago Luchetti and then to the Lajas Valley. The total lake drainage area is about 116.55 km² and is used as a boundary condition in the current model.

According to Flood Insurance Study by Federal Emergency Management Agency [23], the land use on the Río Grande de Añasco watershed are distributed as follows: 278 km² are cropland; 114 km² are pasture; 85 km² are

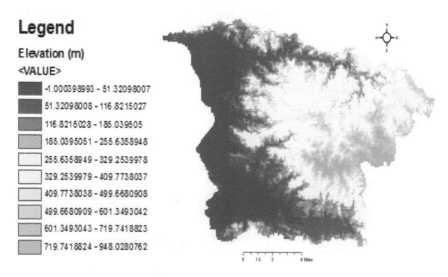

FIGURE 12.2   Digital elevation model (DEM).

forest and woodland; 33 km^2 are idle, and 13 km^2 are urban development and other uses. The vegetation in the floodplain was primary sugar cane. Soils in the floodplain are clay loams. The entire Rio Grande de Añasco watershed is in the humid, mountainous physiographic area of Puerto Rico. The Atalaya Mountains extend from the coastline eastward along the north side of the floodplain, merging with dissected plateau remnants at slightly lower elevations, north of the City of Añasco [23].

Flood problems in this study area are serious and widespread. Periodic flood damage to pastureland, roads, and a number of residential areas is significant. Flood waters have inundated the main Río Grande de Añasco floodplain 17 times in a period of 31 years, an average of approximately once every 2 years. The floodplain of the lower Rio Grande de Añasco has been inundated extensively at least six times during the period 1899–1975: September 1975 (major), September 1928, September 1932, September 1952, October 1970, August 1899, and September 1899 [23].

12.2.2 RÍO GUANAJIBO BASIN

The Río Guanajibo basin originates in the cordillera central of western Puerto Rico. It rises approximately 10 kilometers north-east of Sabana Grande at an elevation of 800 meters approximately. The topography of the area includes mountains, foothills and valleys. The Río Guanajibo valley is approximately 27 km long and is fan-shaped, with a width varying from approximately 0.6 kilometers in the area located between the town of Sabana Grande and San German, to approximately 5.2 kilometers in the Cabo Rojo and Hormigueros region, and approximately 2.8 kilometers in the valley outlet, near the mouth [23]. The Río Guanajibo basin is subdivided into subbasins for each principal tributary: Río Rosario, Río Duey, Río Cain, Río Cupeyes, Río Cruces, and Río Loco. The top of the Guanajibo valley lies in the east of Sabana Grande. In this area, serpentinite and volcanic rocks are predominant, in the south serpentinite predominates in a strip along the border. Rocks along the southern border of the valley near Punta Guanajibo consist of weathered serpentinite, with some volcanic-related rocks.

The urban areas are around Sabana Grande, San German, Cabo Rojo, Hormigueros, and a little portion of the City of Mayagüez. Land use in the Guanajibo River Basin can be divided into three main groups: agriculture with 59%, forested with 33% and residential housing with 8% [23]. Information on the historic floods of the basin can be found in the USGS hydrologic Investigations Atlas HA-456 by Haire [32]. One of the greatest floods ever recorded in the basin was caused by Tropical storm Eloise, which occurred on September 15–17, 1975 and had a recurrence interval of approximately 100 years.

Unfortunately, no efforts have been directed toward obtaining sufficient data to do flow-frequency analyzes. Of the known floods, the events of August 9, 1899, was the largest, followed by the flood of September 13, 1928. Both floods were associated with the passing of a hurricane over the island [32]. Water-surface elevations recovered from these floods were not sufficient to adequately define the floodplain boundaries. Other significant floods occurred on December 3–4, 1960; May 17–18, 1963; July 30, 1963; November 27, 1968; and September 15–17, 1975. The flooding area in this zone has been delineated on the topographic map using the flood of July 30, 1963, it is fairly representative of floods in 1945, 1952, 1954 and 1960 [32].

12.2.3   RÍO YAGÜEZ BASIN

The Río Yagüez Basin is located in the west-central portion of Puerto Rico. It flows westerly into the Bay of Mayagüez. The drainage basin is narrow with a length-width ratio of approximately 10 to 1 and a total drainage area of 35.5 km². The City of Mayagüez, through which Río Yagüez flows, is among the largest cities in Puerto Rico [23]. The largest known flood on Yagüez River occurred on March 3, 1933 24-hours precipitation total of 44.2 centimeters was recorded at Mayagüez by the national Oceanic and Atmospheric Administration (NOAA) on that date. This resulted in a flood with a peak discharge of 708 m³/s and a recurrence interval of 75 years. In 1968, a flood protection project for the City of Mayagüez was started, the total project consisted of a channel and a reservoir to protect the city from floods. Currently, the channel with the existing structures has a capacity of 326 m³/s, but there are plans to rebuild some of these structures, thereby increasing the capacity of the channel [23].

12.3   SOIL CLASSIFICATION

The soil map was provided by United States Department of Agriculture – Natural Resources Conservation service (USDA-NRCS [55]), Soil Survey Geographic Database (SSURGO) for the Mayagüez [89], Lajas Valley [90], Arecibo [91] and Ponce area [92]. These were used in the conceptualization of the soils surface texture for the study area (Figure 3). Hydraulic parameter initial values for clay, loam, clay-loam, gravel, rock and sand soil surface texture were assumed based on values from the literature for representative physical properties of soil texture [82].

The soil textures present in this study as percent of area are clay with 62.49%, clay–loam 24.96%, rock 8.69%, loam 3.00%, sand 0.81% and gravel 0.04%. A soil map describing the class distribution is necessary to assign the values the Green-Ampt infiltration parameters (Figure 12.3).

Harmsen et al. [36] developed an algorithm of Water and Energy Balance for Puerto Rico using data from the Geostationary Operational Environmental Satellite (GOES). GOES-PRWEB uses an energy balance

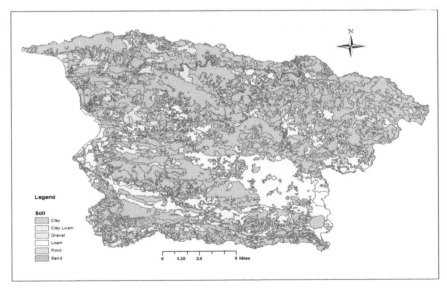

FIGURE 12.3   Soil textures present in the study area [*Source:* Soil Survey Geographic (SSURGO)].

approach similar to Yunhao et al. [109]. The latent heat flux component of the algorithm is used to estimate actual evapotranspiration. The algorithm depends on solar radiation, which is determined using GOES satellite data. Gautier et al. [26] were first to propose a physical model for estimating the incident solar radiation at the surface from the GOES.

Harmsen et al. [36] provided solar radiation data with spatial resolution of one km for Puerto Rico. In this chapter, authors developed a subroutine in MatLab to convert the original one km resolution to 200-meter resolution to obtain potential evapotranspiration estimation in a resolution compatible with the hydrologic model in this chapter.

National Digital Forecast Database [NDFD] estimates daily average wind velocity for Puerto Rico. They adjusted the virtual instrument height, depending on the height of vegetation. Minimum, average and maximum and dew point air temperatures are obtained from a lapse rate approach calibrated for Puerto Rico by Goyal et al. [30]. These temperatures are daily adjusted with a nudging technique, using forecast temperature data from the NDFD [51]. Detailed description of the methodology used to obtain potential evapotranspiration is presented by Harmsen et al. [35, 36].

12.4   LAND USE CLASSIFICATION

A digital map of the land cover developed by the *Xplorah* project [107] was used to conceptualize the different land cover categories present in the study area. The data was developed by the School of Planification of the University of Puerto Rico – Rio Piedras [*Xplorah project*, 107], as shown in Figure 12.4.

Twenty (20) different classes of land cover and forest type are present over the study area corresponding to different kind of forest, woodland and agriculture. The classification of land cover in this model is used to assign values for physical based parameters which are important in the simulation with *Vflo*, other important parameters with the land use are manning's roughness coefficient, rainfall interception, evapotranspiration, crop coefficient and other.

Prieto [61] classified the land use for this watershed in six (6) major categories, shrub land, woodland and shade coffee with an area of 529.16 km², pastures with 172.84 km² of area, urban and barren area with 60.02 km²,

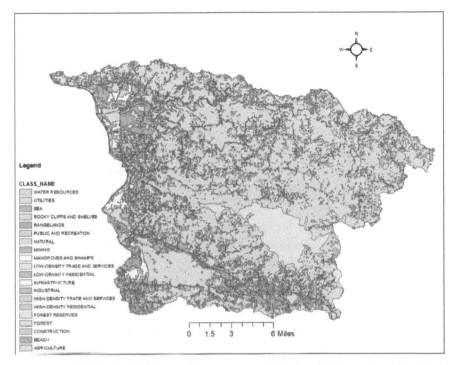

FIGURE 12.4   Land Use by *Xplorah* project [School of Planification of the University of Puerto Rico – Rio Piedras, 107].

agriculture with 55.06 km², other emergent wetlands with 1.26 km² and Quarries, sand and rock with 0.75 km².

12.5   THE LOCAL CLIMATE

The climate of the study area is considered humid subtropical. The average temperature at the Mayagüez City, Puerto Rico station (666073) is 70.7°F between the years 1971–2000, and the average max temperature in the Mayagüez city station between the years 1971–2000 is 88.7°F, National Climatic data Center (NCDC) [52]. The amount of rainfall varies considerably throughout the study area. Most of the rainfall occurs during the month of September with 10.62 inches on average. The months of January through April are considered the dry season with 1.60 inches in January, 2.59 inches in February, 3.35 inches in March and 4.17 inches in April on average rainfall. South-east Regional Climate Center [SERCC, 78] presents detailed reports (figures and tables) on the climate in the study area.

In the west, the sea breeze effect carries wet air from the Mona Channel eastward, converging with the Trade Wind and resulting in intense convective rainstorms almost every afternoon within the MBDB during the wet season. Rainfall and temperature data obtained from the National Climatic data Center [52]. Table 12.1 shows the average temperature between 1948–2012 at Mayagüez City, Puerto Rico, [South-east Regional Climate Center (SERCC), 78] and Table 12.2 presents the precipitation average between 1948–2012 at Mayagüez area [South-east Regional Climate Center (SERCC), 78].

Other record in the Mayagüez area is the station in the Mayagüez Airport. Figure 12.5 shows the average of precipitation monthly between the years 1981 and 2010. This agrees with the Mayagüez city station with September been the month with more precipitation.

12.6   HIGH-RESOLUTION RAINFALL RADAR PRODUCT

Commonly, the flood alert systems have fulfilled the role of providing flood notification to the community and have saved lives and buildings. However, many alert systems fail due to low precision of the models and the sudden change of the atmosphere. One of the greatest sources of uncertainties in the prediction of flooding is the rainfall input [56]. Therefore, it is essential to have an accurate source of rainfall spatial and temporal data, and this is possible with properly working radars.

National Weather Service has a network of approximately 150 Doppler-radar stations S-band (10-cm wavelength) radar distributed across the continental United States, Alaska, Hawaii, Guam and Puerto Rico only one here [52].

The first installation of a WSR-88D for operational use in everyday forecasts was in Sterling, Virginia on June 12, 1992. The radars provide spatial rainfall estimates at approximately 16-km^2 resolution. This network was originally designed to support Departments of Defense, Transportation and Commerce objectives for detection and mitigation of severe weather events [111].

TABLE 12.1 Period record of temperature average monthly between 1948–2012 at climatic station Mayagüez City, Puerto Rico [52]

From Year=1948 To Year=2012

Station:(666073) MAYAGUEZ CITY

Averages Daily Extremes

	Monthly Averages			Daily Extremes				Monthly Extremes				Max. Temp.		Min. Temp.	
	Max.	Min.	Mean	High	Date	Low	Date	Highest Mean	Year	Lowest Mean	Year	>= 90 F	<= 32 F	<= 32 F	<= 0 F
	F	F	F	F	dd/yyyy or yyyymmdd	F	dd/yyyy or yyyymmdd	F	-	F	-	# Days	# Days	# Days	# Days
January	85.1	65.0	75.0	94	20/1985	54	13/1963	81.0	112	72.6	76	1.3	0.0	0.0	0.0
February	85.8	64.5	75.1	96	20/1980	43	02/2004	80.5	112	72.5	75	1.9	0.0	0.0	0.0
March	87.1	65.1	76.1	96	21/1958	56	07/1976	79.9	112	72.4	76	5.5	0.0	0.0	0.0
April	87.9	66.5	77.2	96	14/1951	57	09/1974	80.4	87	74.3	74	7.7	0.0	0.0	0.0
May	89.1	68.4	78.7	98	31/1977	58	02/1975	81.4	87	75.3	100	13.2	0.0	0.0	0.0
June	90.2	69.2	79.7	98	18/1969	60	26/1996	83.1	69	77.0	76	18.4	0.0	0.0	0.0
July	90.3	69.7	80.0	96	30/1970	60	27/1960	82.7	105	77.4	56	21.3	0.0	0.0	0.0
August	90.3	70.1	80.2	98	07/1999	60	13/1994	84.2	99	75.1	96	21.1	0.0	0.0	0.0
September	90.1	70.0	80.1	97	04/1991	61	12/1996	82.6	99	77.0	96	19.8	0.0	0.0	0.0
October	89.4	70.0	79.6	96	27/1999	62	02/1956	83.3	98	77.0	56	16.0	0.0	0.0	0.0
November	87.7	68.5	78.1	95	02/1996	47	07/2004	83.7	98	75.6	54	7.7	0.0	0.0	0.0
December	85.9	66.7	76.3	95	25/1998	58	22/1964	82.8	98	73.8	73	2.3	0.0	0.0	0.0

TABLE 12.1 Continued

From Year=1948 To Year=2012

Station:(666073) MAYAGUEZ CITY

Averages Daily Extremes

	Monthly Averages					Daily Extremes			Monthly Extremes				Max. Temp.		Min. Temp.	
	Max.	Min.	Mean	High	Date	Low	Date	Highest	Year	Lowest	Year	>=	<=	<=	<=
								Mean		Mean		90 F	32 F	32 F	0 F
	F	F	F	F	dd/yyyy or yyyymmdd	F	dd/yyyy or yyyymmdd	F	-	F	-	# Days	# Days	# Days	# Days
Annual	88.2	67.8	78.0	98	19690618	43	20040202	79.3	92	75.8	74	136.3	0.0	0.0	0.0
Winter	85.6	65.4	75.5	96	19800220	43	20040202	80.8	112	73.2	76	5.6	0.0	0.0	0.0
Spring	88.0	66.6	77.3	98	19770531	56	19760307	79.8	87	74.5	74	26.4	0.0	0.0	0.0
Summer	90.3	69.6	80.0	98	19690618	60	19600727	82.3	90	77.6	74	60.8	0.0	0.0	0.0
Fall	89.1	69.5	79.3	97	19910904	47	20041107	82.9	98	76.9	56	43.5	0.0	0.0	0.0

TABLE 12.2 Period record of precipitation monthly average between 1948–2012 at station Mayagüez City, Puerto Rico [52]

From Year=1948 To Year=2012
Station:(666073) MAYAGUEZ CITY
Averages Daily Extremes

	Precipitation											Total Snowfall		
	Mean	High	Year	Low	Year	1 Day Max.	dd/yyyy or yyyymmdd	>= 0.01 in. # Days	>= 0.10 in. # Days	>= 0.50 in. # Days	>= 1.00 in. # Days	Mean	High	Year
	in.	in.	-	in.	-	in.						in.	in.	-
January	1.48	6.26	93	0.01	102	4.60	14/1970	6	3	1	0	0.0	0.0	55
February	1.83	5.99	96	0.01	65	2.47	28/1985	7	4	1	0	0.0	0.0	55
March	2.52	10.29	89	0.00	97	3.96	29/1992	8	4	2	1	0.0	0.0	55
April	4.80	11.89	59	0.41	97	3.94	21/1983	10	7	3	1	0.0	0.0	55
May	7.05	15.00	86	0.60	99	6.80	28/1980	13	10	5	2	0.0	0.0	55
June	6.92	16.11	69	0.84	82	4.20	12/1969	13	10	5	2	0.0	0.0	55
July	8.91	20.90	77	1.00	99	3.90	07/1971	15	12	6	3	0.0	0.0	55
August	9.04	17.47	75	1.25	99	6.68	25/1988	16	13	6	3	0.0	0.0	55
September	10.20	27.02	98	1.25	96	21.30	22/1998	17	13	6	3	0.0	0.0	55
October	8.40	16.18	71	0.51	96	4.03	21/1968	16	12	6	3	0.0	0.0	55
November	4.95	13.75	48	0.88	66	3.39	04/1956	11	8	3	1	0.0	0.0	55
December	2.12	9.57	101	0.03	99	2.68	14/1953	7	4	1	1	0.0	0.0	55

TABLE 12.2 Continued

From Year=1948 To Year=2012
Station:(666073) MAYAGUEZ CITY
Averages Daily Extremes

| | Precipitation | | | | | | | >= 0.01 in. | >= 0.10 in. | >= 0.50 in. | >= 1.00 in. | Total Snowfall | | |
| | Mean | High | Year | Low | Year | 1 Day Max. | dd/yyyy or yyyymmdd | # Days | # Days | # Days | # Days | Mean | High | Year |
	in.	in.	-	in.	-	in.						in.	in.	-
Annual	68.22	94.12	75	16.77	99	21.30	19980922	139	102	45	21	0.0	0.0	55
Winter	5.43	14.29	61	0.38	100	4.60	19700114	20	11	3	1	0.0	0.0	56
Spring	14.37	26.69	80	3.84	97	6.80	19800528	31	22	9	4	0.0	0.0	55
Summer	24.87	41.11	81	7.60	99	6.68	19880825	44	35	17	8	0.0	0.0	55
Fall	23.55	38.72	75	4.27	99	21.30	19980922	44	33	16	7	0.0	0.0	55

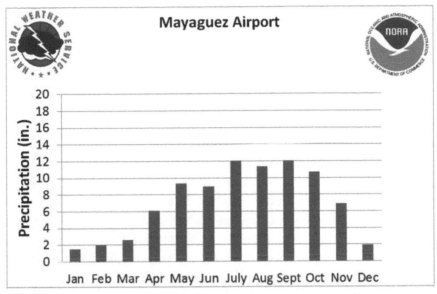

FIGURE 12.5 The average of precipitation recorded for each month of the year between 1981–2010, at Station Mayagüez Airport, Puerto Rico. *Source:* NCDC, (2013). National Climatic Data Center – National Oceanic and Atmospheric Administration (NOAA), US Government. (http://www.ncdc.noaa.gov/)

NEXRAD has been used by the NWS to estimate rainfall in Puerto Rico. The NEXRAD facility for Puerto Rico is located near the City of Cayey at 860 m above mean sea level and at approximately 120 km from Mayagüez city. The location of radars provides full nationwide coverage over the contiguous United States at a specified height above each of the individual radars, but this may present a problem in the western Puerto Rico due to the distance from the NEXRAD radar and topography of the Island. Digital distributed-precipitation radar products can be downloaded directly from NWS.

The WSR-88D (Weather Surveillance Radar 1988, Doppler) radar, commonly referred to as NEXRAD, was developed to replace preDoppler technology radars for the purpose of providing an advanced early warning system for tornadoes. The first prototype system was installed in Norman, Oklahoma, in 1988. The first full scale WSR-88D radar was deployed in 1992. The main objective of the NWS's NEXRAD program from a hydrologist's perspective is to provide, in real-time, accurate quantitative precipitation estimates (QPE) from its network of radars [2].

An equation relating reflectivity (Z) and rainfall (R) as the power function, $Z = aR^b$, is normally used to retrieve estimated values for rainfall rates. The parameters a and b are selected according to the specific region. In Puerto Rico, NWS commonly uses $a = 250$ and $b = 1.2$. The Z-R coefficients have been shown to vary as a function of many factors and previous studies have shown that it is not possible to derive a single equation that is accurate at every point in a given radar domain, and for every storm-type and storm intensity [87]. As part of research in this chapter, it is important develop a program to convert binary coded files into ASCII-formatted files that contain a rainfall intensity estimate in mm/h for every latitude and longitude in the specific area.

The NEXRAD (Next-Generation-Radar) located in Cayey measures reflectivity to one km by one degree resolution for a diameter (distance) of 460 km [52]. Figure 12.6 shows the coverage of NEXRAD radar in Puerto Rico.

Currently, the Puerto Rico Weather Radar Network (PRWRN) administrated by UPR-Mayagüez has five radars; of which three are OTG and two are polarimetric TropiNet (RXM-25) radars.

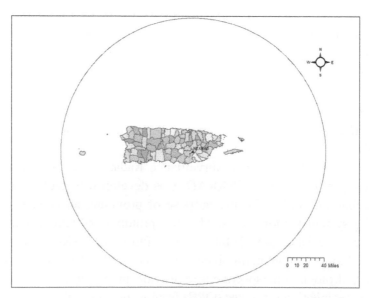

FIGURE 12.6   NEXRAD radar coverage in Puerto Rico.

Figure 12.9 presents the TropiNet radar at Cabo Rojo in Cornelia Hill, while Figure 12.8 shows the TropiNet radar at UPR-Agricultural Experimental Station in Lajas. A new TropiNet radar is being installed at the UPR Agricultural Experiment Station in Isabela, which has the same characteristics as the other two. When all three TropiNet radars are operating simultaneously, the cover area will be approximately one third of the island.

Figure 12.9 shows the coverage of the three TropiNet radars in the western Puerto Rico. The OTG radars were developed with a heterogeneous network using off the shelf hardware. The network was designed to provide detailed precipitation estimates (QPE) to the public, including the NWS staff in Puerto Rico.

12.7 TROPINET RADARS

Radars are active sensors that emit electromagnetic pulses into the surroundings. A typical radar system consists of at least the following four

FIGURE 12.7   TropiNet-1 at Cornelia Hill in Cabo Rojo.

FIGURE 12.8   TropiNet-2 at UPR Agric. Exp. Station in Lajas.

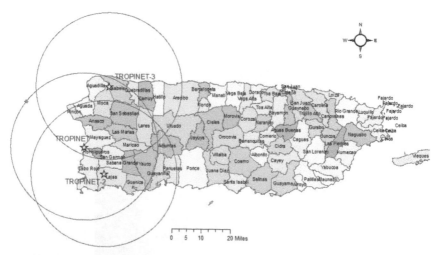

FIGURE 12.9   TropiNet's radars coverage.

components: a transmitter that generates high frequency signals, an antenna that sends the signal out and receives the echoes returned, a receiver that processes the returned signals and a data display systems [67]. Lower frequency and higher wavelength suggest that the radar has robust signal power and less attenuation, the weather radar system discussed in the current research is based in X-band. The common weather radar system can be classified as listed in Table 12.3.

The TropiNet (RXM-25) radars are Doppler polarimetric radars, which allow the radar beam to measure reflectivity close to the ground, overcoming the shadow effect of the Earth's curvature, while maintaining high range and azimuth. The first TropiNet radar has been in operation since February 2012. TropiNet 1 is located in "Cerro Cornelia" Cabo Rojo, Puerto Rico at 18.16°N, 67.17°W, and 200 ft elevation (msl), approximately. The radars, working with the X-band frequency, are about three times stronger than that of the traditional radar frequencies at S-band making the measurements of rainfall more attractive. They have high space and time resolution for weather monitoring and detection, and are capable of generating very high-resolution data with a range of 40 km of radius or maximum radial distance (horizontal range) of 80 km of diameter.

TropiNet radar being Doppler and Polarimetric can show velocity data of the cloud and reflectivity for every azimuth angle from 0° to 12°. TropiNet displays reflectivity logarithmically (10 log(Z)), or dBZ. The working frequency is 9.41 GHz ± 30 MHz, which corresponds to the X-band (in free space has a 3.19 cm wavelength). The TropiNet radar was designed and developed by Colorado State University (CSU) and (UPRM) to serve as the principal Internet-controllable node of the TropiNet radar

TABLE 12.3   Radar Bands with Frequencies and Wavelength. *Source:* [http://stb.ece.uprm.edu/fullscreen/mobile.html]

Radar band	Frequency	Wave length
	GHz	cm
L	1–2	30–15
S	2–4	15–8
C	4–8	8–4
X	8–2	4–2.5
K_U	12–18	2.5–1.7
K	18–27	1.7–1.2
K_a	27–40	1.2–0.75
W	40–300	0.75–0.01

network [25]. The Operational use of radar and hydrological models are indicated in Table 12.4. The Table 12.5 presents the specifications of TropiNet radar.

To analyze the data it was necessary to develop a model to convert raw data to NetCDF data and after convert the reflectivity data in dBZ to rain-rate in (mm/hr) using empirically derived Z-R relationships to transform reflectivity to rain rate. Marshal and Palmer [46] equation is the default Z/R relationship employed by the WSR-88D and TropiNet.

NOAA-NWS [53] report recommended that Z-R relationship in use at the time of the event be changed from $Z = 300R^{1.4}$ to a relationship more representative of raindrop distributions in a warm tropical storm. The Z-R relationship for warm tropical events recommended by the NWS Operational Support Facility since 1995 for all WSR-88D sites experiencing heavy rainfalls, and now adopted by TropiNet is $Z = 250R^{1.2}$ [96].

The Z-R relationship used in Puerto Rico is the convective, furthermore was necessary to define a maximum precipitation rate threshold for decibels above 53 dBZ [96]. The convective rainfall is a type of precipitation with some characteristics like very high horizontal gradient and very

TABLE 12.4 Operational Use of Radar and Hydrological Models [72]

Country	Spatial resolution	Temporal resolution	Radar type	Hydrological model type	Hydrological model name
Czech Republic	2×2 km² (1×1 km² planned)	10 min	C-band	Several	Several including PACK, API Sacramento
Finland	1×1 km²	15	X, C, S-band	Conceptual, distributed	FEI
France	1×1 km²	5 min	C	Conceptual R-R	SOPHIE
Germany	Various projects, resolutions and models				
Poland	1×1 km²	10 min	C-band	Conceptual R-R	IHMS-based
Slovenia	1×1 km²	10 min	C-band	Lumped R-R conceptual	HEC-1
Spain	1×1 km²	6–10 min	C-band	Distributed, grid-based, conceptual	TOPDIST
United Kingdom	Smallest 1×1 km²	5 min	C-band	Various	Various

TABLE 12.5   TropiNet Radar Specifications. *Source:* CRIM, (1998). Center for Municipal Tax Revenues of Puerto Rico.Digital Elevation Model

Transmitter	Specification
Type	**Magnetron**
Center Frequency	9410 +/– 30 MHz
Peak power output	8.0 kW (per channel)
Average power output	12 W (per channel)
Pulse Width	400–660 ns
Polarization	Dual linear, H and V
Max. Duty cycle	0.16%
Antenna and Positioner	**Specification**
Type (diameter)	Dual-polarized parabolic reflector (1.8 m)
3-dB Beam width	1.4 deg
Gain	42 dB
Max. scan rate	60 deg/s
Receiver	**Specification**
Type	Parallel, dual channel, linear I/Q output
Dynamic range	95 dB (BW=1 MHz)
Noise Figure	5 dB
Data Acquisition System	**Specification**
Sampling rate	200 Msps
Dynamic range	105 dB (BW=1 MHz)

large vertical depths. These characteristics imply that the weather radar is the best tool for detecting convective precipitation, but the presence of different types of hydrometeors, especially hail and storm dynamics yielding fast varying Vertical Profile Reflectivity (VPR) usually results in considerable random error in quantitative precipitation estimates. Large differences can be found especially when comparing rain gauges and radar estimates because of the high temporal and spatial variability of the convective storm and related vertical profile of reflectivity [71].

PRWRN has been developing an interactive web site where it is possible to observe weather conditions in real time using both, the OTG and TropiNet radars. It is possible to observe the overlap between these radars and NEXRAD. Figure 12.10 presents the web site under development.

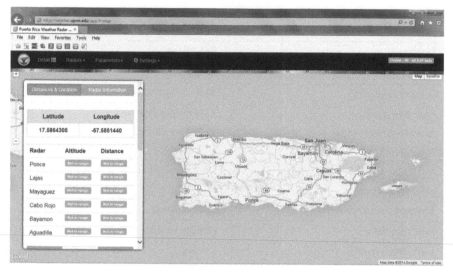

FIGURE 12.10   Coverage Website OTG's and TropiNet radars in real time, [http://stb.
ece.uprm.edu/fullscreen/mobile.html].

The web site is user friendly and accessible to the interested public who
wish to observe weather conditions in real time with higher resolution
than NEXRAD. This web site includes five radars: TropiNet – Cabo Rojo,
TropiNet-Lajas, OTG-Mayagüez- OTG-Ponce, and OTG-Aguadilla. Only
one TropiNet-Cabo Rojo data was used in this research.

On the other hand, Luz Torres-Molina with support of Red de Radars
del Tiempo project developed a Rain Gauge network for comparison of
data radar from TropiNet-Cabo Rojo. These rain gauges series are distributed at University of Puerto Rico Mayagüez Campus (UPRM) and nearby
locations.

12.8   RADAR DATA PROCESSING TROPINET

A radar application in MatLab was developed to access the store of binary
volume files that contain the respective information as determined by the
operator like reflectivity, azimuth, velocity, beam width, range, elevation
and other radar products. The operator can apply one of several possible
scan configurations. For instance, in the Range Height Indicator (RHI), the
radar holds its azimuth angle constant but varies its elevation angles. This

is essential to provide vertical resolution where the radar continuously scans through elevation angles at a given azimuth angle (Figure 12.11). Another common scan configuration is the Plan Position Indicator (PPI). The radar holds its elevation angle constant but varies its azimuth angle, rotating through 360 degrees (Figure 12.12).

For this research, it was necessary to hold the radar scan in PPI with a constant elevation angle of 3 degrees. Every radar scan has two angles of 3 degrees and 5 degrees with a duration time of 30 seconds. The data information is saved in the server at http://www.weather.uprm.edu. The

FIGURE 12.11   Range Height Indicator (RHI).

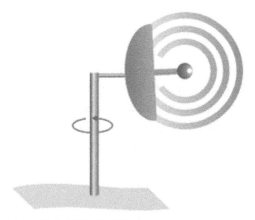

FIGURE 12.12   Plan Position Indicator (PPI).

raw data files are stored by date, every hour, minute and second of scan in binary format. Each volume scan from radar has been interpolated to a fixed Polar grid and, after it is necessary, to convert to the fixed Cartesian grid. As part of the effort to further post-process the radar data, a model in MatLab was developed. This model performs the conversion from raw data polar coordinate system to ASCII data to Geographic coordinate system necessary for the hydrological software, *Vflo*.

In addition, a comparison between NEXRAD and TropiNet in random pixels was made with the objective of validating the rainfall location using a time series for every storm in each pixel.

12.8.1   RADAR DATA PROCESSING: NEXRAD

The NOAA webpage (http://www.ncdc.noaa.gov/nexradinv/map.jsp) indicates the data from NEXRAD. NEXRAD inventory has the option to choose day and product [52]. There are a total of 41 level III products routinely available from the National Climatic Data Center (NCDC), general products include the baseline reflectivity, velocity and algorithmic graph products spectrum width. The base reflectivity [N0R] product is used to detect precipitation, evaluate storm structure, locate boundaries and determine hail potential, and a display of echo intensity measured in dBZ. Four lowest elevation angles are available. For this study, Level III [N0R] short-range base reflectivity (16 level/230 km) with 0.5 degrees was used.

The WSR-88D NEXRAD radar data is stored on the NCDC robotic mass storage system, commonly known as the Hierarchical data Storage System (HDSS). The data is easily accessible with the NEXRAD Inventory Search tool, which allows users to view the data completeness and download individual products. The ordered data is ready for use with the NCDC Weather and Climate Toolkit. Each order may contain up to 24 h of data at a time for a single site. Once the data is downloaded, it is necessary to change data format from NetCDF to ASCII. This is only possible with a developed routine in MatLab from the current this research.

12.9   RAIN GAUGE NETWORK

As leverage to the NSF – CASA center, with support from NOAA's Cooperative Remote Sensing Science and Technology Center (CREST), a rain gauge network was deployed for validation of data from NEXRAD, OTG and TropiNet radars. The rain gauges are distributed over the University of Puerto Rico at Mayagüez Campus (UPRM) and other locations close to the campus.

These rain gauges are tipping bucket-type rain gauges that measure rainfall in 0.254 mm (1/100th inch) increments. The self-emptying, tipping bucket design is accurate (±2%) and reliable. The logger is capable of saving 48 days of rainfall data with a 10 min reading interval. Double rain gauges were installed at each location to minimize errors in data collection.

A major source of error in hydrologic models is the poor quantification of the areal distribution of rainfall, typically due to the low density of rain gauges. For a good spatial distribution of data it is necessary put hundreds of rain gauges in a small area, otherwise it is not possible to obtain a good precipitation distribution.

Some data from TropiNet radars was compared with rain gauge data for selected storms. Figure 12.13 shows the distribution of the rain gauge network in the vicinity of UPRM campus.

Rainfall dates are traditionally obtained from an often-sparse network of rain gauges that may not record the rainfall event with adequate spatial and temporal scales, especially for heavy convective storms when significant rainfall occurs over a limited areal extent [83]. Weather radar has enormous potential in this field, as it can measure rainfall in real-time with high spatial resolution and temporal continuity [83].

FIGURE 12.13   Detailed rain gauges network.

A favorable rainfall distribution is only acquired with radars, therefore it is necessary the use of weather radars, a rain gauge located at a single point may not represent an extensive area, with only one value. The spatial distribution of precipitation can have a major influence on the hydrological models Errors may occur in the resulting hydrograph when the spatial pattern of the rainfall is not preserved. These errors will be magnified for intense, short duration and localized events especially in areas of high topographic variability subject to convective storms [105].

Similarly, errors in rain gauges are known from turbulence and increased winds around the gauge, affecting precipitation quantification in events where the wind is an important factor (e.g., hurricanes). Investigators have used mean areal precipitation as calculated by, for example, Thiessen polygons, [95, 105], and interpolation methods, such as Spline, Inverse Distance Weights, and Krigging and polynomial surface. But all of these methods are limited by the number of rain gauges [105].

12.10   PHYSICALLY-BASED HYDROLOGIC MODEL

The hydrologic model used in this research is *Vflo* [97]. *Vflo* is a fully distributed physically based hydrologic (PBD) model capable of using geographic information and multisensory input to simulate rainfall runoff from major river basins to small catchments (Figure 12.14).

Vflo is a hydraulic approach to hydrologic analysis and prediction. Overland flow and channels are simulated using the Kinematic Wave Analogy (KWA). The model uses GIS grids to represent the spatial variability of factor controlling runoff. Runoff production is from infiltration excess and is routed downstream using kinematic wave analogy. Computational efficiency of the fully distributed physics-based model is achieved using finite elements in space and finite difference in time. *Vflo* is suited for distributed hydrologic forecasting in post-analysis and in a continuous operation mode, derives its parameters from soil properties, Land use, and topography and in this case the precipitation is obtained from radar TropiNet. The goal of distributed modeling is to better represent the spatial-temporal characteristics of a watershed governing the transformation of rainfall into runoff.

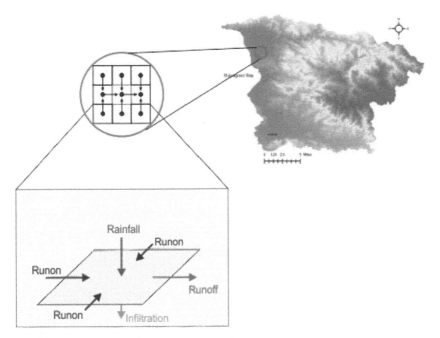

FIGURE 12.14   Detailed GIS grid runoff in the watershed.

The hallmark of *Vflo* is prediction of flow rates and stages for every grid cell in a catchment, watershed, river basin or region. *Vflo* provides high-resolution, physics-based distributed hydrologic modeling for managing water from catchment to river basin scale. Improved hydrologic modeling capitalizes on access to high-resolution quantitative precipitation estimates from model forecasts, radar, satellite, rain gauges, or combinations of multi sensor products.

Model input consists of rain-rate maps at any time interval from radar or multisensor sources. Data input for this model (besides rainfall), is derived from various commonly available sources of digital data. Parameters include topography and drainage networks derived from a digital elevation model (DEM), infiltration derived from soils, and hydraulic roughness derived from land use/cover. These parameters may be input and edited manually or via ArcView grids.

The model formulation is a kinematic wave analogy (KWA) for overland flow is a simplification of the conservation of mass and momentum equations, wherein the principle gradient is the land surface slope. The conservative form of the full dynamic equation relates the temporal and x-direction gradients of flow depth y and velocity V as:

$$\frac{\partial V}{\partial t} + V\frac{\partial v}{\partial x} + g\frac{\partial y}{\partial x} - \left(s_0 + s_f\right) = 0 \tag{1}$$

where, if all other terms are small or of an order of magnitude less than the bed slope s_o, or friction gradient, s_f then the KWA is an appropriate representation of the wave movement downstream [10], V is the component of velocity g, is acceleration due to gravity, $(\partial V/\partial t)$ is local acceleration, $V(\partial v/\partial x)$ is horizontal momentum advection and $(\partial y/\partial x)$ is hydrostatic pressure. The one-dimensional continuity equation for overland flow, with depth h, resulting from rainfall excess is:

$$\frac{\partial h}{\partial t} + \frac{\partial(uh)}{\partial x} = R - I \tag{2}$$

where, R is rainfall rate; I is infiltration rate; h is flow depth and u is overland flow velocity.

In the KWA, the bed slope is associated with the friction gradient which amounts to the uniform flow assumption. Using this fact together with an appropriate relationship between overland flow velocity u and flow depth h such as the Manning equation is obtained:

$$u = \frac{s_0^{1/2}}{n} h^{2/3} \tag{3}$$

where, s_0 is the bed longitudinal slope and n is the Manning's hydraulic roughness.

Velocity and flow depth depend on the land surface slope and the friction induced by the hydraulic roughness. Important parameters are the saturated hydraulic conductivity K controlling infiltration rate I, and Manning's roughness n are three of the most important parameters within the model. Hydraulic conductivity controls the total amount of water that will be partitioned into the surface runoff and the subsurface, whereas the hydraulic roughness mainly affects the peak flow and the time to peak [98]. Model results obtained from Eqs. (1)–(3) are adjusted by scalars applied to spatially distributed parameters:

$$\frac{\partial h}{\partial t} + \varpi \frac{s^{1/2}}{n} \frac{\partial h^{5/3}}{\partial x} = \gamma R - \varrho I \tag{4}$$

where, the three scalars ϱ, γ, ϖ and are multipliers controlling the infiltration rate I rainfall rate R, and hydraulic roughness n, respectively. The flow depth is h, and slope s, is the principal land-surface slope at the center of each grid cell.

The slope and hydraulic roughness are spatially variable, while rainfall, infiltration and flow depth are spatially and temporally variable. Infiltration excess (IE) is treated by the model as the source of runoff. The model represents overland flow as a uniform depth over a computational element. From hillslope to stream channel, there may be areas of IE and Saturation Excess (SE), however the model treat runoff generation as solely IE. Simulation of IE requires soil properties and initial soil moisture conditions. The well-known Green-Ampt equation is used to account for the effects of initial degree of saturation on infiltration rate [98].

12.10.1   CALIBRATION PROCESS

There is a sequence called the "Ordered Physics Based Parameters Adjustment" (OPPA) method developed by Vieux and Moreda [100]. The calibration process (OPPA) approach include estimates of the spatially distributed parameters from physical properties, assigns channel hydraulic properties based on measured cross-sections where available, studies model sensitivity for the particular watershed, and identifies response sensitivity to each parameter. It furthermore runs the model for a range of storm from small, medium to large events. It observes the characteristics of the hydrograph over the range of storm size and any consistent volume bias. Then it derives a range of response for a given change in a parameter and categorizes and ranks parameter sensitivity according to response magnitude.

The optimum parameter is that set which minimizes the respective objective function and matches volume by adjusting hydraulic conductivity. It can match the peak by adjusting overland flow roughness and readjust hydraulic conductivity and hydraulic roughness if necessary. The *Vflo* model does not simulate base flow, only direct runoff; it can be simulated assigning a fixed value to every channel cell for every event to simulate.

For a long-term analysis it is necessary to quantify the base flow using known methodologies [31]. The OPPA procedure outlined above can be stated as: increasing the volume of the hydrograph is achieved by decreasing hydraulic conductivity, and similarly, increasing peak flow is achieved by decreasing hydraulic roughness.

12.11   INUNDATION (FLOOD) MODEL

The Inundation Analyst extension is a *Vflo* [97] that provides images, animations and simulated inundation, which is an indication of flood risk. The extension is especially useful for flood management applications. For example, a forecast inundation is useful for operational decisions, warning or notification, and coordinating emergency response. The Inundation Analyst operates independently from the *Vflo* model, but can use data exported from *Vflo* as input for generating inundation forecasts. The Inundation Analyst requires a digital elevation model (DEM), a flow direction map, a channel flow direction map, and stage data. All input data must be in ESRI ASCII grid format (*.asc*). The DEM and flow direction maps must have the same number of columns and rows. The DEM must be in units of either feet or meters, in this research the units are in meters. Stage data inputs are exported from a *Vflo* model. The resolution of the maps affects the quality of inundated area display, so high-resolution data are recommended. When a flow direction map of a different grid definition is used, filter files called BAG files (*.bag*) may be used to convert *Vflo* stage data to the grid definition of the flow direction map. Background images can be included at any resolution, so long as their extent is the same the other form the DEM and flow direction map. Background images must be in JPEG or bitmap format [103].

Some storms were used as validation of the flow/inundation model. The methodology of validation included: comparing the stream flow and stage using gauge data from the U.S. Geological Survey (Current Water Data for Puerto Rico, [95]) with the observer data from TropiNet radar and rainfall nowcasting. All input data are ASCII and the flow direction is extracted from the DEM watershed. The DEM have units of meters, the stage data input are exported from *Vflo* model, and a background watershed image is included in bitmap format. The inundation results are listed in order to create the animation. Once all stage files are listed in the appropriate order, the images that are produced show the primary inundation Analyst window.

12.12   STOCHASTIC MODELING OF RAINFALL OF SHORT-TERM DURATION

For atmospherics phenomena, it is difficult to predict deterministically what will occur in the future. A mathematical expression which describes the probability structure of the time series that was observed due to the phenomenon is referred to as a stochastic process. The precipitation is an example of stochastic phenomenon that evolves in time according to probabilistic laws. A time series model is adapted to a series in order to calibrate the parameters of stochastic process. Stochastic models are able to provide reliable predictions over small temporal and spatial scales, which are interested in hydrological applications.

Other types of prediction are the meteorological models, they produce qualitative and quantitative rainfall forecasting for 24–72 h. At these forecasting horizons, an absolute precision is not required, but rather an order of magnitude. They are based on atmospheric phenomena developing on a synoptic scale, but in general they are not able to provide reliable predictions for small temporal and spatial scales, which are of interest in hydrological applications [14].

12.12.1   TIME SERIES ANALYSIS

A time series is a set of observations that are arranged chronologically. In time series analysis, the order of occurrence of the observations is crucial. When a meteorologist wants to predict a storm or a rainfall using forecasting or nowcasting, the more important factor is the chronological order of the data or the data time series. If this chronologic data is ignored, the information contained will be difficult to use.

In the time series analysis, stochastic models are used for describing the system hydrology for purposes that include modeling, forecasting, generating and investigating the underlying characteristics of the rainfall data. A time series is a set of observations that are arranged chronologically.

In this work, the observations are reflectivities which were changed to rainfall. They were derived from TropiNet radar obtained between the months of March and December of 2012 and some months in 2014. Some of the precipitation events from TropiNet radar observed were modeled to obtain the nowcasting of 10 min, 20 min and 30 min, and then this data was compared with the observed data of 10 min, 20 min and 30 min. In total 10 events in Range Height Indicator (RHI) mode were used between 2012 and 2014, when the radar was available.

12.12.2   TYPES OF FORECASTS

There are some properties needed to distinguish between different types of forecast. Forecast can extend to different scales in space and time; the spatial is doing reference in a fixed location in a specific area o city, e.g., the precipitation on a grid from TropiNet radar over Mayagüez city. The temporal range of a forecast is furthermore called lead-time. Short range forecast cover very close events, like the next few hours or next minutes as our case, the long range forecast is considered the mean value of a meteorological parameter over a few days or months.

In this research, the data is correlated in space and time, where the strength in general decreases with spatial and temporal distance. Our models are designed to do forecast in time and space. This increases the difficulty as compared with prediction models that only use the forecast in time at a given place (e.g., forecast in rain gauges).

Other types of forecast are deterministic. In this case a single forecast value is issued at each occasion, pretending a confidence that hides the forecaster's uncertainty about the outcome. They are easy to interpret even for user without stochastic background knowledge. The simplest case is a deterministic binary forecast. This area decision, like yes or no, and additionally a generalization in the forecast if necessary, distinguishes between types of variables to be forecasted. The variable of interest can be ordinal, which can be expressed by a number and can be defined by an appropriate number of threshold values (e.g., light rain, middle rain or heavy rain).

Other variable of interest is the nominal, where there is no natural ordering, like qualitative observation of the kind of precipitation (e.g., snow, rain, ice or other). A deterministic evaluation is furthermore named Quantitative Precipitation Forecast (QPF), which induces the user to suppress information and judgment about uncertainty. In fact, it may create the illusion of certainty, while a probabilistic forecast is indicated as Probabilistic Quantitative Precipitation Forecast (PQPF). In order to reflect the uncertainty of the future outcome, probabilistic statements are more appropriate.

For this research a methodology that embrace a space-time stochastic model is used, and is considered a "discrete time-series model" that include a special kind of nonlinear model with stochastic and deterministic components. Here, the rainfall process is described at a discrete time steps, are not intermittent and, therefore, can be applied for describing the forecast within storm rainfall.

The other investigators prefer to use of meteorological model. These are useful qualitative and quantitative rainfall forecasting tools on 24–72 h interval and on a large spatial scale. In such cases, indeed absolute precision is not required for practical application. In meteorological models when the forecasting lag time and spatial scale decrease, the effectiveness and the precision of kind of model additionally decrease [43]. The next section shows some types of forecast models widely used.

12.13   AUTOREGRESSIVE-MOVING-AVERAGE MODELS

Autoregressive-moving-average models (ARMA) are mathematical models of autocorrelation in a time series. ARMA models are widely used in hydrology and were popularized by Box and Jenkins [5] who elaborated a comprehensive theoretical and practical development of time series models. There are several possible reasons for fitting ARMA models to data. ARMA modeling can contribute to understanding the physical system by revealing something about the physical process that builds persistence into the series. ARMA models can additionally be used to predict behavior of a time series from past values alone. Such a prediction can be used as a baseline to evaluate possible importance of other variables to the system.

The model consists of two parts: an autoregressive (AR) part and a moving average (MA) part. The AR model expresses a time series as a linear function of its past values. The order of the AR model indicates how many lagged values are included. The MA model is a form of ARMA model in which the time series is regarded as a moving average of a random shock. The model is usually then referred to as the ARMA (p,q) model where p is the order of the autoregressive part and q is the order of the moving average part. ARMA models in general, after choosing p and q, are fitted by iterative procedure of a nonlinear least squares regression to find the values of the parameters which minimize the error term. The ARMA modeling process is commonly an iterative, trial and error process. Thus, it is necessary to use the least possible number of parameters that will adequately produce forecasted values with similar statics of the historical data [19].

ARMA is a methodology widely used to do predictions of all types, for economy as well as for the weather predictions. In any case, it is necessary to have a long historical data. In the literature ARMA model has been used to predict at one or two rain gauges at a single point but not at radar field. Since the ARMA model predicts at a single point. This is an important reason to avoid the use of ARMA methods in this research. This principle was applied to this thesis or this model, at the same time the principle of parsimony to obtain results in the model with small possible error.

12.14   POINT PROCESS MODEL

Point Process is a type of random process for which any action consists of a set of isolated points in time or in space. The example more global in point process model is the Poisson Process that counts the number of events (storm) and the time that these events occurs in a given time interval,. Usually the time between each events development has an exponential distribution and the numbers of occurrences are independent of each event (storm).

The Point process model has been used commonly to forecast rainfall in which storm origins occur in a Poisson process. The Point process model is applied at a single site or fixed point where the storms arrive in a Poisson process. Each storm incorporates a group of random number of rain cell, where each cell has a random duration or lifetime and depth. The total rate of precipitation at time (t) is the sum of contributions from all active cells at (t) [70]. This type of model uses complex equations and the analysis of precipitation is in time at a fixed point in space and the properties of the natural process can be deduced via the mathematical model.

Stern and Coe [84] have modeled daily rainfall in which wet and dry days occur in a Markov chain with seasonally dependent transition probabilities. In it, the amounts of rain per wet day have a gamma distribution with seasonally dependent parameters.

12.15 SPECIAL "NONLINEAR EMPIRICAL MODEL"

An algorithm for predicting 10, 20 and 30 min in advance the spatial distribution of rainfall rate is based on the assumption that TropiNet radar rainfall rate data provides estimations of the rainfall with high spatial and temporal resolution. Some researchers have compared radar rainfall data with rain gauge measurements [6, 71, 77, 108]. These comparisons may not been useful since a rain gauge measures precipitation at a single point located at the surface level, whereas the weather radar measures the average of reflectivity at certain elevation and over a much larger area. A stochastic function is used to estimate the rainfall rate based on reflectivity.

When a rain gauge is compared with radar, it is expected that the average rainfall will behave as an individual point. It is known that the average will behave differently than that of an individual observation; therefore, these quantities should not be expected to be equal. When several rain gauges are averaged and compared with the radar measurements, the average of the rain gauges is inconsistent because it was developed with few points whereas the average of the radar was developed with a much larger number of points. The rainfall modeled over a watershed shows that the peak flow measurements and overall runoff from radar performed better that the estimated peak flow using rain gauges [76]. Additional studies have concluded that the peak discharge of stream-flow computed with radar data were more accurate than those computed from rain gauges alone [68]. Thus, there is no instrument that precisely measures the amount of rainfall over a large area. The weather radar provides an estimation of the rainfall rate over larger areas.

The suggested algorithm uses TropiNet (RXM-25) data to predict the variability of the rainfall field in time and space. It is assumed that for a short time period (10, 20 and 30 min), a rain cloud behaves as a rigid object, with all pixels moving in the same direction at a constant speed. Thus, the most likely future rainfall areas are estimated by tracking rain cell centroid advection in consecutive radar images. The suggested algorithm is a special kind of nonlinear model with stochastic and deterministic components. The rainfall process exhibits significant changes in time and space, and it can be characterized as a nonstationary stochastic process. To face the nonstationary characteristic of the process, parameters are estimated at every time and spatial domain.

The model consists in considering the rainfall shape data as a rectangular grid with 940 columns and 740 rows of pixels for a total of 695,600 pixels, every pixel size is 0.06 kilometers wide and 0.06 kilometers long. From the grid data select a zone of 81 pixels that was divided in squares of $\Delta x \times \Delta y$ pixels, where (Δx) is referenced to columns of 9 pixels and (Δy) rows of 9 pixels with total zones of 8528 (82×104) in every window (Figure 12.15). Several zones sizes were explored for Δx and $\Delta y = \{7, 9, 11, ..., 25\}$ and it was found that the larger the zone size, the larger the number of degree of freedom. However, resolution was degraded with increased zone size.

In the model, the use of the same zone in the before windows ($t - 1$) ($t - 2$) and is necessary (Figure 12.16). Every zone (9×9) should have

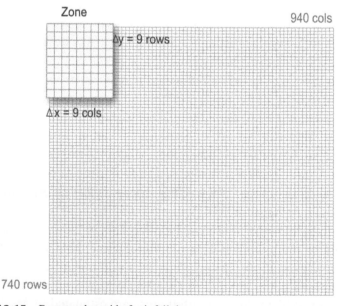

FIGURE 12.15   Rectangular grid of rainfall data.

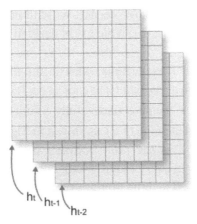

FIGURE 12.16   Zone 9 × 9 at time t, t–1, and t–2.

a minimum of 24 rain pixels with 20 degrees of freedom. Zones with less pixel of rain could not be selected to forecast analysis. In zones where the prediction movement suggest there is a rainfall cell but the zone has not the necessary pixels required (24 pixels) an interpolation was applied. The interpolation was *"Kriging simple"* using the 20 five pixels nearest to pixel that has no prognostic.

The model is defined by the following equation:

$$h_{t,k(i,j)} = \{\alpha_{t,k} + \left(\beta_{t,k} - \alpha_{t,k}\right)\Phi_{t,k}$$
$$\left[1 - e^{-\sum\left(\delta 1_{t,k}\,\bar{h}_{t-1,k(i,j)} + \delta 2_{t,k}\,\bar{h}_{t-2,k(i,j)} + \delta 3_{t,k}\,Z_{t-1,k(i,j)}\right)}\right] + \varepsilon_{t,k(i,j)}\} \tag{5}$$

where, (i,j) represents the geographic position or coordinates latitude and longitude of every pixel in the grid, k is the zone. This process starts in pixel 1 until pixel 8528. In every zone, unknown parameters should be determined (α, β, Φ, $\delta 1$, $\delta 2$, $\delta 3$): α is the minimum value found between previous values of $h_{t-1,k(i,j)}$ and $h_{t-2,k(i,j)}$ in their respective zones (k), β is the reflectivity maximum value found between previous values of $h_{t-1,k(i,j)}$ and $h_{t-2,k(i,j)}$ in the specific zone (k).

The mathematical structure of the model is based on a previous work by Ramírez-Beltran [63]. In the current work, this model was used because this scheme ensures that rainfall forecasts will fall inside of the most likely rainfall intensity domain [α, b], which was derived by the observed local rainfall distribution.

$\overline{h}_{t-1,k(i,j)}$ is the reflectivity average value in the time $(t-1)$. The average value was determined in every pixel into each zone. It was obtained averaging the eight pixels closest to the pixel under study. Similarly, $\overline{h}_{t-2,k(i,j)}$ is the average reflectivity value in the time $(t-2)$, as shown in Figures 12.17 and 12.18.

The variable $Z_{t-1,k(i,j)}$ is the ratio between the pixels with maximum reflectivity. $Z_{max(t-1),k(i,j)}$ in every cloud or cell and the nearby pixels $Z_{i(t-1),k(i,j)}$ forming the cloud or cell and the random variable $e_{j,k(i,j)}$ is a sequence of an unobserved random variable with mean zero and constant variance associated to the pixel (i,j). Therefore, we have:

$$N_r = \frac{Z_i - Z_{min}}{Z_{max} - Z_{min}} \tag{6}$$

The variable Phi ($\Phi_{t,k}$) is changing in the equation every zone (9×9) in each window. This variable was determined first by linearization of the

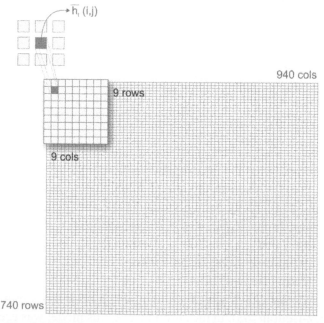

FIGURE 12.17   Average pixels at a specific zone using the eight nearest pixels.

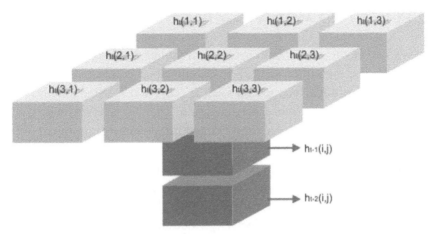

FIGURE 12.18 Average pixel at h_t for (2,2).

nonlinear equation (Phi-initial) and after using optimization nonlinear techniques with constrains *Sequential Quadratic Programming* (SQP), where the Phi parameter is a bias correction factor and its maximum value must not exceed 1.1.

$$0 < \Phi_{t\,opt}^{k} \leq 1.1 \tag{7}$$

The initial coefficient deltas (δ_1, δ_2, and δ_3) were obtained through the estimation method "*least squares*" by linearization of nonlinear equation (exponential). Once the variables initial deltas were found, the next step is to find the variable Phi ($\Phi_{t,k}$) initial. These values were used to forecast rainfall at one lead-time and successively with the following forecasts.

An additional important step in this research was defining the cloud motion vector in each cell, with capacity to predict the rainy pixel areas, plus the joint with the forecast rainfall estimation using the main equation. For the cloud motion, h_{t-1} and h_{t-2} direction and movement were necessary to determine the direction motion vector and velocity. This velocity is compared with velocity obtained for TropiNet to assure the right movement of the clouds.

The proposed rainfall prediction algorithm requires the implementation of three major tasks: (1) Develop the cloud motion vector, (2) Predict the future location of the rainy pixels, and (3) Estimate the rainfall rate in the future rainy pixels.

12.15.1 CLOUD MOTION VECTOR

Derivation of the cloud motion vector requires tracking cloud rainfall cells [64]. The algorithm identifies first the cloud convective core based in a sequence of radar images between h_{t-2}, h_{t-1} and h_t using an empirical distribution method for cloud classification. Then by determining the distance between cloud center at time $t-1$ and the cloud center at time t of the same cloud (Figure 12.19).

The motion algorithm was based on a spatial and temporal comparison, classifying clouds with high reflectivity and removing pixel with very low reflectivity, in this work the minimum reflectivity was 3 dBZ. The next step is the normalization of reflectivity values between a range of zero and one using minimum and maximum values of reflectivity in each image or windows, as shown in the following equation, where N_r is the normalized reflectivity, Z_i reflectivity in each pixel, Z_{min} minimum reflectivity 3 dBZ and Z_{max} is the maximum reflectivity in the window.

$$N_r = \frac{Z_i - Z_{min}}{Z_{max} - Z_{min}} \tag{8}$$

The classification of the normalized values is divided into two groups. This result was stored in a binary matrix B_r. The value N_r exceeding the percent of pixel with a minimum reflectivity $N_{r,min}$ is assigned value of one and the value N_r that is smaller than the percent of pixel with a minimum reflectivity $N_{r,min}$ is assigned the value of zero. In this case, $N_{r,min}$ is 10 percent of pixels with values of minimum reflectivity.

$$B_r = 0 \text{ if } N_r < N_{r,min} \tag{9}$$

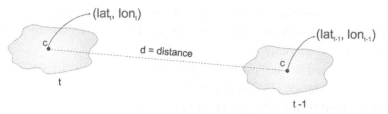

FIGURE 12.19   The motion cloud between time $t-1$ and time t.

$$B_r = 1 \text{ if } N_r < N_{r,min} \qquad (10)$$

Next, the binary matrix is imported into another subroutine which classified the cell with separations by rows and columns, grouping the continuous pixels. The method for cloud classification looks for a minimum group of 250 pixels successive with binary data. When there are more than three (3) rows or three (3) columns of pixels without data into the grid it is possible to have a division of clouds. This is the form to separate every one cell or cloud (Figure 12.20).

The contiguous pixels in the radar image are used to form the convective cell. It is necessary to know the centroid of every cell and the latitude and longitude of each pixel into the cell at the times $t-2$, $t-1$, and t.

The distance (d), direction (θ) and velocity (v) are properties between the centroids of the cells that are moving in every lag-time. This is calculated using the next equations.

$$di = \sqrt[4]{\left(x_{ti} - x_{ti-1}\right)^2 + \left(y_{ti} - y_{ti-1}\right)^2} = \sqrt[2]{\Delta x^2 + \Delta y^2}\,(km) \qquad (11)$$

$$\theta = \tan^{-1}\frac{\Delta y}{\Delta x}\,(rad) \qquad (12)$$

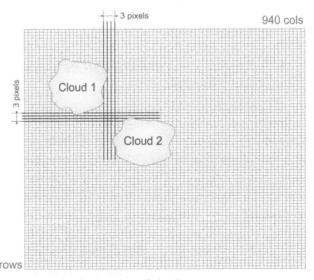

FIGURE 12.20   An example of separation of clouds.

$$v = \frac{d\,(km)}{t\,(min)} \qquad (13)$$

To determine the centroid of the cells, it is necessary to calculate latitude $(\overline{La})$ and longitude $(\overline{Lon})$ of every pixel group.

$$\overline{La} = \frac{1}{n}\sum_{i=1}^{n} la_i \qquad (14)$$

$$\overline{Lon} = \frac{1}{n}\sum_{i=1}^{n} lo_i \qquad (15)$$

Dixon and Wiener [20] found that a convective cell have a mean velocity of 64 km/hr. This value agrees with the velocity cell measure from other research [54]. For this model a velocity means of 72 km/h approximately or 12 km/(10 min) was used. To apply this maximum distance between clouds at every lag time of 200 pixels was necessary if the analysis is every 10 min. If this analysis time increases, then the distance could increase (Figure 12.20).

The 200 pixels represent the maximum distance of translation cell in two different times. Figure 12.21 shows the cloud 1 moving a Δt from $(t-2)$ to $(t-1)$. This is furthermore referred to as *coverage diameter* in two successive times or a *delta time* Δt.

12.15.2 ESTIMATION OF PRECIPITATION USING THE NOWCASTING MODEL

The precipitation was estimated using Eq. (5), applied to each zone in every time window. The rain estimated, $\hat{h}_{t+1,k(i,j)}$ at time $(t-2)$ is the result of the prediction interval Δt (10, 20, 30) between the instants $i\Delta t$ and $(i+1)\,\Delta t$. It is a function of the previous database on dynamic sets of parameters. The constants (α, β, Φ, $\delta 1$, $\delta 2$, $\delta 3$) were determined in each zone (9×9) using optimization techniques for nonlinear regression equations. The main equation includes four fundamental products: $\overline{h}_{t-1,k(i,j)}, \overline{h}_{t-2,k(i,j)}, h_{t-1,k(i,j)}$ and $Z_{t-1,k(i,j)}$. These are the average observed rain at time $t-1$ and $t-2$. The average is calculated between the eight nearest

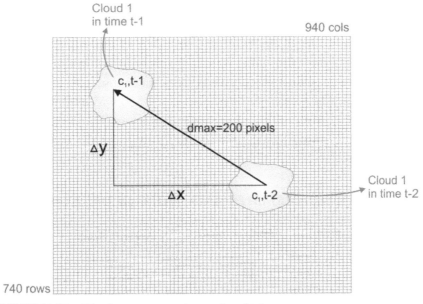

FIGURE 12.21 Cloud movement at time $t-2$ and $t-1$.

pixel to the prediction pixel. The other $h_{t-1,k(i,j)}$ is the value of the rain at time $(t-1)$, and the $Z_{t-1,k(i,j)}$ is the ratio of reflectivity at time $(t-1)$. The Eq. (5) has some restrictions in the parameters (α, β, Φ, $\delta 1$, $\delta 2$, $\delta 3$), which are changing in time and space. The clouds are in movement and the values of the variables are changing at every time and space domain. After the optimization, the deltas values are restricted to be positive or equal to zero.

$$\delta_{i,t,k} \geq 0; \ i = 1, 2, 3 \tag{16}$$

The variables of α and β are the minimum and maximum reflectivity values, respectively, between the last two windows at $(t-1)$ and $(t-2)$ at the zone (9×9), these variables are changing in time and space (every zone 9×9). Moreover, the variable $\Phi_{t,k}$ changes in every zone and time windows but having a restriction limit of 1.1 in the optimization routine.

$$\alpha = \min\left(Z_{t-1}, Z_{t-2}\right) \tag{17}$$

$$\beta = \max\left(Z_{t-1}, Z_{t-2}\right) \tag{18}$$

$$0 < \Phi_{t,k} \le 1.1 \qquad\qquad (19)$$

Once the variables were found, the next step was to estimate the rain rate forecast in every pixel using Eq. (5). Pixels for which it was not possible to do the estimation prediction or there is not enough data at time $t - 1$ and/or at $t - 2$. The "*Kriging*" interpolation method was used to estimate the rain pixel to derive the corresponding predictors [110]. Figure 12.22 shows the cloud movement sequence with the centroid and their distance between them.

12.15.3 INITIAL VARIABLES AND THEIR OPTIMIZATION

The variables into the nonlinear equation model are fundamental in the precipitation forecast trend. A well-planned approach is needed to properly solve the nonlinear constrained problem. The explored approach includes two steps: (i) identifying the initial point and (ii) using a constrained nonlinear optimization technique to estimate the final parameter set for each zone and every window [63].

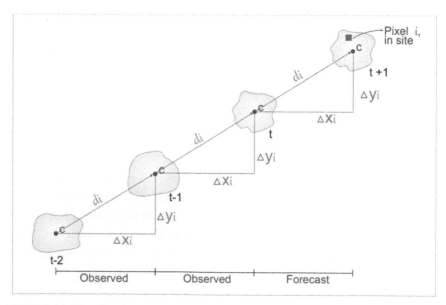

FIGURE 12.22 Cloud movement sequence.

To estimate the initial values of deltas, it was not necessary to apply the constrain, so that the initial deltas values can be positives or negatives. The Eq. (5) was linearized by considering values of $\overline{h}_{t-1,k(i,j)}$, $\alpha_{t,k}$, $\beta_{t,k}$ and $Z_{t-1,k(i,j)}$ and the unknown values of $\delta1_{t,k}$, $\delta2_{t,k}$, $\delta3_{t,k}$ left the parameter phi $\Phi_{t,k}$ temporarily ignored.

This method consists in solving the equivalent linear model and using these values as the initial point. The convergence of nonlinear routine heavily depends on the selections of the initial points. Thus, if the initial point is far away from the optimal solutions the algorithm may converge to a suboptimal point or may not converge. Linearizing the Eq. (5) and ignoring the phi variable [63], we get:

$$-Ln\left[1-\left(\frac{h_{t,k(i,j)}-\alpha_{t,k}}{\beta_{t,k}-\alpha_{t,k}}\right)\right]=\sum\left(\begin{array}{c}\delta1_{t,k}\overline{h}_{t-1,k(i,j)}+\delta2_{t,k}\overline{h}_{t-2,k(i,j)}\\+\delta3_{t,k}Z_{t-1,k(i,j)}\end{array}\right)+\varepsilon_{t,k(i,j)} \quad (20)$$

where

$$\beta_{t,k}>h_{t,k(i,j)} \ and \ \alpha_{t,k}<h_{t,k(i,j)} \quad (21)$$

$\varepsilon_{t,k(i,j)}$ is an unknown random variable at time t and at location (i,j) of the k zone. The initial values of delta are obtained by solving the linear regression Eq. (20) by the least square method.

The phi parameter is a bias correction factor and can be estimated using a second linear regression. Once the delta values are estimated, the next step is to find the phi value $(\Phi_{t,k})$, which can be estimated using the following equation:

$$\left[\frac{h_{t,k(i,j)}-\alpha_{t,k}}{\beta_{t,k}-\alpha_{t,k}}\right]=\Phi_{t,k}[1-e^{-\Sigma(\delta1_{t,k}\overline{h}_{t-1,k(i,j)}+\delta2_{t,k}\overline{h}_{t-2,k(i,j)}+\delta3_{t,k}Z_{t-1,k(i,j)})}]+\varepsilon_{t,k(i,j)} \quad (22)$$

For: $\beta_{t,k}>h_{t,k(i,j)}$ and $\alpha_{t,k}<h_{t,k(i,j)}$

$\varepsilon_{t,k(i,j)}$ is an unknown random variable at time t and at location (i,j) in the zone 9×9.

Simplifying Eq. (22) with the initial delta estimates, the following equation is obtained:

$$\lambda_{t,k(i,j)} = \Phi_{t,k}\left(\theta_{t,k(i,j)}\right) + \eta_{t,k(i,j)} \qquad (23)$$

where,

$$\lambda_{t,k(i,j)} = \frac{h_{t,k(i,j)} - \alpha_{t,k}}{\beta_{t,k} - \alpha_{t,k}} \qquad (24)$$

$$\theta_{t,k(i,j)} = \left[1 - e^{-\Sigma \bar{\delta} 1_{t,k} \bar{h}_{t-1,k(i,j)} + \bar{\delta} 2_{t,k} \bar{h}_{t-2,k(i,j)} + \bar{\delta} 3_{t,k} Z_{t-1,k(i,j)}}\right] \qquad (25)$$

$\eta_{t,k(i,j)}$ is an unknown random variable in Eq. (23) at time t and at location (i,j) of the k zone, $\bar{\delta}$'s are the previous estimated or initial values of deltas.

The next step is to find the optimum values of variables $\delta 1_{t,k}$, $\delta 2_{t,k}$, $\delta 3_{t,k}$, and $\Phi_{t,k}$ from initial values determined in the previous steps. The parameters of the nonlinear regression model can be easily estimated by solving a constrained nonlinear optimization problem. Since the main model or Eq. (5) includes four parameters with a bounded constraint:

$$\delta i_{t,k} \geq 0; \ i = 1, 2, 3 \qquad (26)$$

$$0 < \Phi_{t,k} \leq 1.1 \qquad (27)$$

Therefore, it can be solved by using the *sequential quadratic programming* algorithm [49, 66]. The derived initial point was ingested into the constrained nonlinear subroutine to facilitate convergence. The parameters of the exponential term were restricted to be positive, and the phi parameter was restricted to be in the range of 0 to 1.1. This threshold was derived by inspection and using statistical analysis. The optimization objective was to minimize the errors between the estimate values for the regression and the observed values by radar.

In these regions during the prediction, there are clouds (or cells) present in the movement estimation, but not the required minimum number of pixels. The pixels estimation predictions were obtained by *Kriging* interpolation.

12.15.4 LEAST SQUARE METHOD

The least squares estimate of the multiples regression parameters were used to calculate the initial values of deltas variables. The multiple linear regression model is typically stated in the following form:

$$y_{\jmath} = U_0 + U_1 x_{1\jmath} + U_2 x_{2\jmath} + \dots + U_N x_{N\jmath} + \epsilon_{\jmath} \qquad (28)$$

Where, the dependent variable is $y_{\jmath}$, U_0, U_1, U_0, ... U_N are the regression coefficients and $\varepsilon_{\jmath}$ is the random error assuming $E(\varepsilon_{\jmath}) = 0$ and $Var(\varepsilon_{\jmath}) = \sigma^2$ for $\jmath = 1, 2, \dots M$.

The multiple linear model can be expressed in matrix format:

$$y = XU + \varepsilon, \text{ where} \qquad (29)$$

$$X = \begin{bmatrix} 1 & x_{12} \dots & x_{1N} \\ x_{21} & x_{22} \dots & x_{2N} \\ x_{m1} & x_{m2} \dots & x_{MN} \end{bmatrix} U = \begin{bmatrix} U_0 \\ U_1 \dots \\ U_{N-1} \end{bmatrix} c = \begin{bmatrix} \epsilon_1 \\ \epsilon_2 \dots \\ \epsilon_M \end{bmatrix} \qquad (30)$$

And finally U values are estimated solving the following multiple linear regressions equation:

$$U = (X' X)^{-1} X' y \qquad (31)$$

It was assumed that $(X' X)$ is a nonsingular matrix [106].

12.15.5 SEQUENCE QUADRATIC PROGRAMMING

The function used for optimization was *fmincon*. This function has a constrained minimum of a scalar function of several variables starting at an initial estimate. This is generally referred to as constrained nonlinear optimization or nonlinear programming [49].

The function *fmincon* uses one of four algorithms: *active-set, interior-point, sqp or trust-region-reflective*. The Sequential Quadratic Programming (SQP) is one of the most successful methods for the numerical solutions of constrained *nonlinear optimization problems* (NLP) [4].

A nonlinear programming problem is the minimization of a nonlinear objective function $f(\mathrm{x})$, $\mathrm{x} \in \mathbb{R}^{m}$ of m variables, subject to equation and inequality constrains involving a vector of nonlinear functions (x). The formulation can be:

$$
\begin{aligned}
&\textit{minimize } f(\mathrm{x}),\ \mathrm{x} \in \mathbb{R}^{m} \\
&\textit{subject to } \mathbf{g}(\mathrm{x}) \leq 0\ i = 1,2,\ldots,mm \\
&\qquad\qquad \mathbf{h}(\mathrm{x}) = 0
\end{aligned}
\tag{32}
$$

where, $f \colon \mathbb{R}^{m} \to \mathbb{R}$ is the objective functional, the functions $\mathbf{h} \colon \mathbb{R}^{m} \to \mathbb{R}^{mm}$ and $\mathbf{g} \colon \mathbb{R}^{m} \to \mathbb{R}^{p}$ describe the equality and inequality constraints.

The nonlinear optimization problem (NLP) has special cases linear and quadratic programming routines, when f is linear or quadratic and the constraint functions $\mathbf{g}$ and $\mathbf{h}$ are affine. SQP is an iterative routine, which models the NLP for a given iterative x^{k+} by a Quadratic Programming (QP) subroutine, solves that QP sub problem, and then uses the solution to construct a new iterative x^{k+1}. This construction is done in such a way that the sequence (x^{k}) converges to a local minimum x^{*} of the NLP.

The NLP resembles the Newton and quasi-Newton methods for the numerical solution of nonlinear algebraic systems of equations. However, the presence of constraints renders both the analysis and the implementation of SQP methods much more complicated [37].

12.15.6 KRIGING INTERPOLATION

Kriging is based on the assumption that the parameter being interpolated can be treated as a regionalized variable. A regionalized variable is intermediate between a truly random variable and a completely deterministic variable in that it varies in a continuous manner from one location to the next. Therefore, the points are near to each other and have a certain degree of spatial correlation. Yet, points that are widely separated are statistically independent [13].

The *Kriging* techniques are based on the estimation of weighting coefficients with an assumption of unbiased-ness. Each data has its own coefficient w_{i}, which represent the influence of a particular data on the value of the final estimation at the selected grid node. The relationship between the existing data and the estimation point has been expressed by *variogram* values or by covariance in case of second order stationarity. Such values describe the spatial dependence and the influence of the particular location in terms of its distance and direction from the estimated location [45]. The basic equation used in ordinary *Kriging* is as follows:

$$F(x, \eta) = \sum_{i=1}^{n} w_i f_i \tag{33}$$

where, n is the number of scatter points in the set, f_i are the values of the scatter points, and w_i are the weights assigned to each scatter point. The weights are found through the solution of the simultaneous equations:

$$w_1 S\left(d_{11}\right) + w_2 S\left(d_{12}\right) + w_3 S\left(d_{13}\right) + \rangle = S\left(d_{1p}\right)$$
$$w_1 S\left(d_{12}\right) + w_2 S\left(d_{22}\right) + w_3 S\left(d_{23}\right) + \rangle = S\left(d_{2p}\right) \tag{34}$$
$$w_1 S\left(d_{13}\right) + w_2 S\left(d_{23}\right) + w_3 S\left(d_{33}\right) + \rangle = S\left(d_{3p}\right)$$

where, $S(d_{ij})$ is the model variogram evaluated at a distance equal to the distance between points i and j. It is necessary that the weights sum to unity.

$$w_1 + w_2 + w_3 = 1.0 \tag{35}$$

The *Kriging* techniques add some constraints to the matrices, to minimize the error, and these techniques are unbiased-ness estimations. These factors would describe some external limit on the input data, which cannot simply be observed in the measured values [45]. The constraint factor in Ordinary *Kriging* equations is called the Lagrange multiplicator (Λ). It is used to minimize possible estimation error and then the Eq. (34) can be written as:

$$w_1 S\left(d_{11}\right) + w_2 S\left(d_{12}\right) + w_3 S\left(d_{13}\right) + \rangle = S\left(d_{1p}\right)$$
$$w_1 S\left(d_{12}\right) + w_2 S\left(d_{22}\right) + w_3 S\left(d_{23}\right) + \rangle = S\left(d_{2p}\right) \tag{36}$$
$$w_1 S\left(d_{13}\right) + w_2 S\left(d_{23}\right) + w_3 S\left(d_{33}\right) + \rangle = S\left(d_{3p}\right)$$

where:

$$w_1 + w_2 + w_3 + 0 = 1.0 \tag{37}$$

The equations are then solved for the weights w_1, w_2, and w_3. The f value of the interpolation point is then calculated as:

$$f_p + w_1 f_1 + w_2 f_2 + w_3 f_3 \tag{38}$$

An important feature of *Kriging* is that the variogram can be used to calculate the expected error of estimation (σ^2) at each interpolation point since the estimation error is a function of the distance to surrounding scatter points. The calculation of error variance for the output pixel estimate includes adding the Lagrange coefficient:

$$\sigma_2 = w_1 S(d_{1P}) + w_1 S(d_{1P}) + w_1 S(d_{1P}) + \Lambda \tag{37}$$

12.16   SELECTION OF EVENTS

To select the events, it was necessary to analyze every storm during 2012 and 2014 that was collected by the radars. The analysis has three steps: The first was taking every minute data from TropiNet radar and plot it. For this, it was necessary to create an efficient routine in MatLab to determine that the radar data had not interruptions or was damaged. If the radar had corrupt data, the storm is discarded. In some cases, it was found that the radar collected data in *Plan Position Indicator* (PPI) and after the radar is changed to *Range High Indicator* (RHI), such data was also discarded.

The next step was to select the radar data with the same elevation angle (3°). The TropiNet radar has the capacity of store data with two o more different elevations angles. Within the model it was necessary to include a subroutine with efficiency to select a determined elevation angle. The final step was to choose those precipitations that have data with complete storm duration.

Table 12.6 includes the dates and specifications of every storm in the current study. The information incorporated in the column "Storm Impact" was provided by NWS at Carolina, Puerto Rico [58].

TABLE 12.6   Characteristics of Studied Storms

Date	Duration (UTC)	Storm type	Storm impacts at western Puerto Rico
March 28, 2012	7 hr. 16:27–23:58	Stationary trough	Impacts rivers, water on the road, and significant rainfall accumulation
March 29, 2012	6 hr. 00:36–06:53	Stationary trough	Impacts rivers, water on the road, significant rainfall accumulation
April 30, 2012	5 hr. 17:55–22:21	Convective storm	Numerous showers over western Puerto Rico at the afternoon
October 10, 2012	5 hr. 16:10–21:43	Convective storm	Some urban flooding
February 12, 2014	7 hr. 16:00–23:29	Heavy convective storm	Reduced visibilities and ponding of water on roadways and low lying areas
May 06, 2014	7 hr. 16:45–23:59	Convective storm	Street flooding and reduced visibility on the highways.
May 21, 2014	7 hr. 16:46–23:00	Heavy convective storm	The water covers the roadway. Ponding of water on roadways
June 29, 2014	5 hr. 17:00–22:00	Convective storm	The shower activity produced periods of moderate to locally downpours
June 30, 2014	4 hr. 16:00–20:15	Thunderstorms associated to the leading edge of a tropical wave	Moderate to heavy rain, urban and small stream flood advisory
July 05, 2014	4 hr. 16:44–20:00	Convective storm	Heavy rain, urban flood.

12.17   HYDROLOGIC MODEL COMPOSITION

As mentioned in this chapter, the hydrological model used in this research was *Vflo*. This model uses finite elements, which can simulate streamflow based on geospatial data to simulate interior locations in the drainage network and determine channel flow and overland flow.

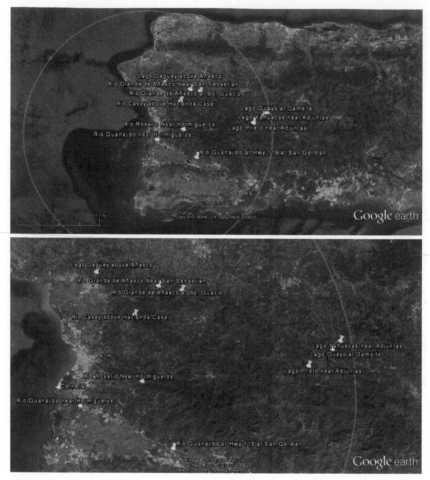

FIGURES 12.23 AND 12.24   Flow stations within radar coverage (top) and within basin (bottom).

Figures 12.23 and 12.24. Flow stations within radar coverage (top) and within basin (bottom).

It was fundamental to study the physical configuration of the watershed, such as a Digital Elevation Model (DEM), the digitized topography, soils map, land use map and information about the basin. Some hydrologic and hydraulic studies have been conducted by Sepulveda et al. [1996]; Villalta [103]; Prieto [61]; Rojas [69]. In addition, other studies by U.S Geological Survey -Current Water Data for Puerto Rico [2014] and FEMA [23] are used in this research as additional information. Some stations from the

TABLE 12.7   USGS Flow Stations

Source	ID Station	Station Name	Lat.	Long.	Elev. (m)	Data
USGS	50131990	Rio Guanajibo at Hwy 119 at San German	18.09	−67.03	45.0	Rain, Stage
USGS	50136400	Rio Rosario near Hormigueros	18.17	−67.07	50.0	Rain, Stage, Flow
USGS	50138000	Rio Guanajibo near Hormigeros	18.14	−67.15	2.2	Rain, Stage, Flow
USGS	50141500	Lago Guayo at Damsite near Castaner	18.21	−66.83	426.8	Rain, Stage
USGS	50142500	Lago Prieto near Adjuntas	18.19	−66.86	600.2	Rain, Stage
USGS	50146073	Lago Daguey above Añasco	18.301	−67.13	40.0	Rain, Stage
USGS	50141100	Lago Yahuecas near Adjuntas	18.22	−66.82	426.8	Rain, Stage
USGS	50143930	Rio Grande de Añasco at Bo. Guacio	18.28	−67.02	64.9	Rain, Stage
USGS	50144000	Rio Grande de Añasco near San Sebastián	18.285	−67.05	31.6	Rain, Stage, Flow
USGS	50145395	Rio Casey above Hacienda Casey	18.25	−67.08	75.0	Rain, Stage, Flow

* U.S Geological Survey – Current Water Data for Puerto Rico, 93.

USGS were used to compare and validate the runoff with the results from the hydrological model using radar data (Table 12.7, Figures 12.23, 12.24).

FEMA [23] implemented the most recent *Flood Insurance Study* (FIS) for the Commonwealth of Puerto Rico. Standard hydrologic and hydraulic study methods were used to determine the flood hazard data required for this countywide FIS. The flood events have magnitude of exceeding once at any given day during the recurrence period of 10 years, 50 years, 100 years and 500 years. These events have a percent chance of 10%, 2%, 1% and 0.2%, respectively. The equation employed were *Mean Annual Rainfall* (MAR) obtained from *Mean Annual Precipitation* (MAP) developed by NOAA in 2006 [precipitation record 1971–2000]. The regression analysis was performed based on depth to rock (DR) and contributing drainage area (CDA) as variables that govern the peak streamflow. A summary of

TABLE 12.8 Drainage Area Peak Discharge Relationship. *Source:* FEMA. Flood insurance study: Preliminary for Commonwealth of Puerto Rico: June 22. Federal Emergency Management Agency (FEMA), US Government;22 June 2012; Volume 1 of 5

Drainage area (sq. km)	Station name	Peak discharge (m³/s)			
		10 year	50 year	100 year	500 year
467.73	Rio Grande Añasco at Mouth	1,809	3,797	5,130	10,542
347.33	Rio Grande Añasco Near San Sebastián	1,390	3,031	4,078	8,329
385.26	Rio Grande Añasco upstream confluence Rio Casey	1,527	3,289	4,432	9,070
414.88	Rio Grande Añasco downstream confluence Rio Casey	1,631	3,481	4,695	9,624
35.4	Rio Yagüez at Mouth	292	595	770	1,289
329.65	Rio Guanajibo at Mouth	1,352	3,896	5,745	14,294
310.53	Rio Guanajibo Near Hormigueros	1,215	3,637	5,343	13,196
91.39	Rio Guanajibo at Hwy 119 at San German	604	1,325	1,713	2,991
303.04	Rio Guanajibo downstream confluence Rio Rosario	1,206	3,507	5,137	12,620

drainage area-peak discharge relationship for all of the streams studied is shown in Table 12.8 [23].

The following sections present the analysis of each variable in the hydrological model and determine the best parameters for a good operation. The analysis was based on existing literature within the study area.

12.17.1 POTENTIAL EVAPOTRANSPIRATION

A GOES satellite-based potential evapotranspiration (PET) product, with resolution of 1 kilometer over the entire island each day, was used in this research. The PET product was obtained from Dr. Eric W. Harmsen of the Agricultural and Biosystems Engineering Department, UPRM [59].

Although PET method by Hargreaves [33, 34] is simpler to use, yet it does not yield PET that can be close to actual field conditions.

One of the most used methods to calculate PET or reference evapotranspiration, and the method used in this study, is the FAO Penman-Monteith method [22]. A large number of empirical methods have been developed over the last 50 years, and the Penman-Monteith method was considered to offer the best result with minimum possible error. The Penman-Monteith reference evapotranspiration equation is given by

$$ETo = \frac{0.408\Delta\left(R_n - G\right) + \gamma \dfrac{900}{T + 273} u_2\left(e_s - e_a\right)}{\Delta + \gamma\left(1 + 0.34u_2\right)} \tag{38}$$

where, ETo is reference evapotranspiration (mm day^{-1}), R_n net is radiation at the crop surface (MJm^{-2}day^{-1}), G is soil heat flux density (MJm^{-2}day^{-1}), T is mean daily air temperature at 2 m height ($^\circ$C), u_2 is wind speed at 2 m height (ms^{-1}), e_s is saturation vapor pressure (kPa), e_a is actual vapor pressure (kPa), $e_s - e_a$ is saturation vapor pressure deficit (kPa), Δ is slope vapor pressure curve (kPa$^\circ$C^{-1}), γ is psychrometric constant (kPa$^\circ$C^{-1}).

The Eq. (40) uses standard climatological records of solar radiation, air temperature ($^\circ$C), humidity and wind speed (ms^{-1}). The weather measurement should be made at 2 m (or converted to that height) above an extensive surface of a hypothetical green grass with an assumed height of 0.12 m, a fixed surface resistance of 70 sec m^{-1} and an albedo of 0.23.

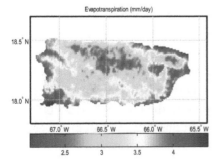
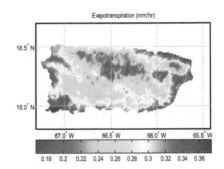

FIGURE 12.25 Left panel potential evapotranspiration (mm/day) and right panel potential evapotranspiration (mm/h) on March 28, 2012, Puerto Rico.

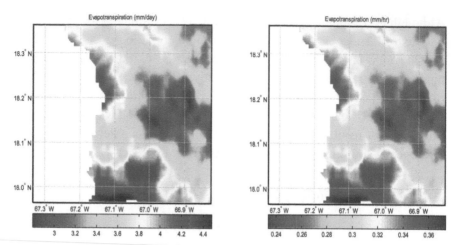

FIGURE 12.26 Left panel potential evapotranspiration (mm/day) and right panel potential evapotranspiration (mm/h) in the basin area on March 28, 2012.

The original PET data resolution is 1 km and the units are (mm/day). The hydrological *Vflo* model uses PET in units of mm/h and the same resolution as the Digital Elevation Map in the current study. A subroutine in MatLab was developed to change the resolution of the PET data from 1 km to 200 meters, and the units from (mm/day) to (mm/hours).

The current study's days were: March 28, 2012; March 29, 2012; April 30, 2012; October 02, 2012; February 02, 2014; May 06, 2014; May 21, 2014; June 29, 2014; June 30, 2014 and July 05, 2014. Figures 12.25 and 12.26 show the potential or reference evapotranspiration for March 28, 2012.

To evaluate the study area in western PR, it was necessary to develop an algorithm in MatLab to extract values of PET from the PR datasets and to assign them to appropriate locations within the study area (Figure 12.26). Finally, the 1-km PET data was projected onto a 200-meter resolution grid using an interpolation methodology in MatLab. Interpolation is a method for estimating the value at a query location that lies within the domain of a set of sample data points. This transformation was successful for the ten storm days analyzed and the data provided by the GOES-based PET was more accurate than PET based on a limited number of available weather stations (two) within the study area.

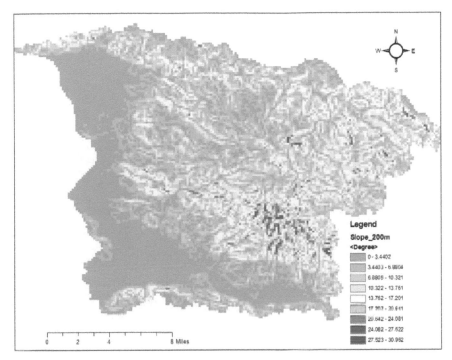

FIGURE 12.27 Basin slope map 200 m resolution.

12.17.2 SLOPE CHARACTERISTICS

The slope map was developed using the digital elevation map (DEM) at 200 meters and 10 meters resolution from USGS. The digital elevation model (DEM) data consist of a sampled array of regularly spaced elevation values referenced horizontally either to a Universal Transverse Mercator (UTM) projection or to a geographic coordinate system. The grid cells are spaced at regular intervals along south to north profiles which are ordered from west to east. Figure 12.27 presents the slope map for the basin derived from the DEM at 200 meters resolution.

An aspect map is elaborated in Figure 12.28. The aspect map is a measured counterclockwise in degree from 0 (due north) to 360 (again due north, coming full circle). The value of each cell in an aspect grid indicates the direction in which the cell's slope faces. Flat slope have no direction and are given a value of −1. There are many different reasons to use the

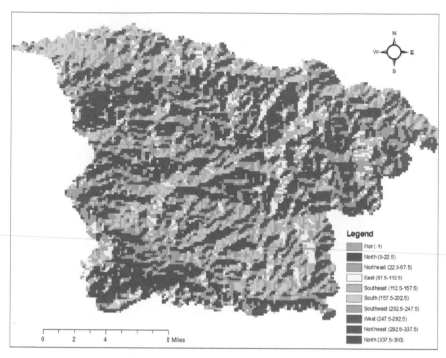

FIGURE 12.28   Basin aspect map 200 m resolution.

aspect function. For example it can be used to identify areas of flat land, slope in a mountainous region, and locations where is possible identify the runoff direction.

12.17.3   CHANNEL SPECIFICATIONS

The study area includes three main rivers and their branch, Rio Grande de Añasco, Rio Guanajibo and Rio Yagüez, (Figure 12.29).

The roughness coefficients developed by FEMA [23] gives a general roughness for the Rio Añasco of 0.040 in the channel and 0.100 in the overbank. In the Rio Yagüez, the roughness range is between 0.030 to 0.050 in the channel and for the overbank it is between 0.150 and 0.200. For the Rio Guanajibo, the channel roughness coefficient for the channel is between 0.040 and 0.045 and the overbank is 0.100.

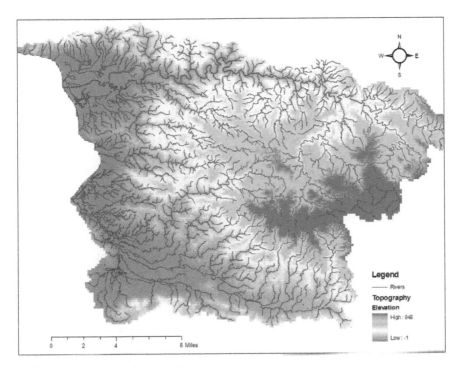

FIGURE 12.29 Map showing three rivers.

TABLE 12.9 Surveyed Sections Coordinates at Rio Grande de Añasco [103]

Sections	Location Coordinates UTM, NAD 1927		Average reach from the mouth (km)
	Latitude	**Longitude**	
P1	2,019,561.15	721,233.41	53.80
AN12	2,019,437.49	721,056.76	52.48
ANCO	2,021,257.06	717,276.30	46.24
ANC2	2,021,592.71	717,240.97	45.99
ANC1	2,021,575.04	716,905.32	45.43
AMA	2,020,603.43	714,785.44	41.592
AN21	2,021,098.07	714,538.12	40.82
AN22	2,021,469.05	714,379.13	40.41
AN32	2,021,981.35	713,319.19	39.08
AREA	2,022,246.34	710,792.99	36.32

TABLE 12.9   Continued

Sections	Location Coordinates UTM, NAD 1927		Average reach from the mouth (km)
	Latitude	Longitude	
AREABA	2,022,317.00	710, 510.34	35.99
GRAVERO ANTES	2,023,465.27	707,895.92	30.26
GRAVERO DESPUES	2,023,500.60	707,295.19	29.56
AN40	2,023,694.93	706,765.22	28.51
ANCG	2,022,617.32	704,044.70	21.79
ANCG2	2,021,769.37	701,730.50	16.60
ESPINO ANTES	2,022,264.00	699.504.62	14.19
ESPINO DESPUÉS 1	2,022,122.68	699,363.29	13.99
ESPINO DESPUÉS 2	2,021,274.73	699,151.31	12.74
OVEJAS-LILLY 1	2,021,398.39	697,932.37	11.17
OVEJAS-LILLY 3	2,020,635.42	696,879.87	8.53
SECCIÓN K	2,021,342.50	695,800.43	5.86
SECCIÓN L	2,021,512.95	694,840.73	3.49
SECCIÓN N	2,020,963.71	693,294.01	1.78
SECCIÓN O	2,020,824.82	692,574.31	1.07

TABLE 12.10   Surveyed Sections Coordinates at Rio Guanajibo [103]

Sections	Location Coordinates UTM, NAD 1927		Average reach from the mouth (km)
	Latitude	Longitude	
ANTES 114	2,006,663.29	695,231.54	6.69
S. DESPUÉS 114	2,006,763.13	696,049.47	6.45
S. DESPUÉS 102	2,006,933.46	695,802.80	6.05
SECCIÓN S-3	2,007,315.21	695,321.20	4.71
SECCIÓN S-2	2,008,184.45	694,528.31	3.22
SECCIÓN S1	2,008,237.31	694,287.51	2.81
SECCIÓN S2	2,008,848.12	693,541.61	1.57
SECCIÓN S3	2,009,053.68	693,300.81	1.19
SECCIÓN S4	2,009,359.09	693,054.14	0.81
SECCIÓN S5	2,009,756.40	692,713.49	0.46

Using ArcGIS, three necessary products were determined to include in the hydrologic model *Vflo*. These are flow direction, overland slope and stream location, the products were developed with an extension of ArcGIS "Arc Hydro" using a DEM of 30 meters from the USGS, other cross section were obtained using DEM of 10 meters where no data was available to define the flood plain in these areas, and channel slope. In most rivers section

TABLE 12.11   Surveyed Sections Coordinates at Rio Yagüez [103]

Sections	Location Coordinates UTM, NAD 1927		Average reach from the mouth (km)
	Latitude	Longitude	
SECCIÓN 1	2,014,456.14	699,585.62	6.77
SECCIÓN 2	2,014,340.71	699,545.94	6.42
SECCIÓN 3	2,014,346.72	699,175.58	5.90
SECCIÓN 4	2,014,213.24	698,919.45	5.23

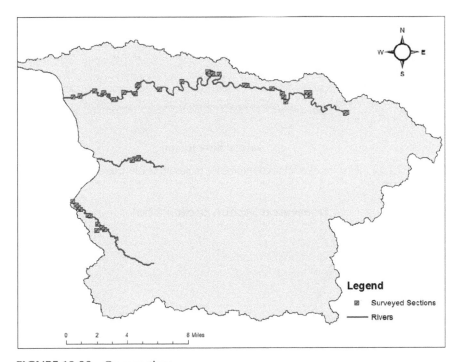

FIGURE 12.30   Cross sections.

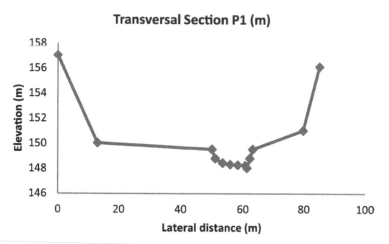

FIGURE 12.31   First transversal section farthest to mouth at Rio Grande de Añasco.

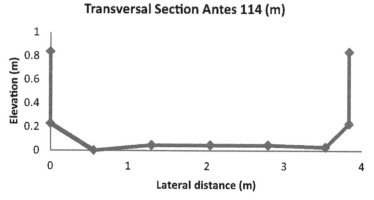

FIGURE 12.32   First transversal section farthest to mouth at Rio Guanajibo.

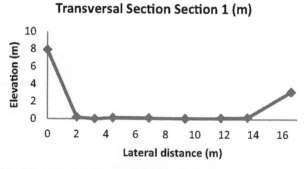

FIGURE 12.33   First transversal section farthest to mouth at Rio Yagüez.

channel width is about 5 to 10 meters, coinciding with Rojas [69]. Villalta's [103] survey sections data was provide by Alejandra Rojas [69], we can observe in Table 12.9 the surveyed sections of Rio Grande de Añasco.

Table 12.10 show the surveyed sections of Rio Guanajibo and Table 12.11 presents the surveyed sections of Rio Yagüez, the sections conserve the original name present in Villalta [103], Prieto [61] and Rojas [69].

Figure 12.30 presents the first upstream surveyed sections in the study area. Figure 12.31 presents the first upstream surveyed sections in Rio Grande de Añasco. Figure 12.32 shows the first upstream surveyed section in Rio Guanajibo and Figure 12.33 presents the first upstream surveyed section in Rio Yagüez. The others transversal sections were also included into the hydrologic model.

12.17.4   INFILTRATION AND ROUGHNESS PARAMETERS

The infiltration is an important parameter to be able to estimate the run-off. The runoff is caused only when the rainfall rates exceed infiltration rates. The hydrologic model use Green-Ampt infiltration routine to model infiltration. Other characteristic parameters in the infiltration process are necessary: Hydraulic conductivity, wetting front, effective porosity, soil depth, initial saturation, abstraction and impervious area, these variables are affected by land use and soils properties. The infiltration parameter was developed using the SSURGO maps and database from USDA [89–92], which contains six textural soil classes in the basin area.

Figure 12.34 presents the six basic textures into the basin area, a large amount area of clay is observed. The soils name present into the clay area are: Alluvial land, Aguilita, Aibonito, Bajura, Consumo, Daguey, Delicias, Humatas, Lares, Jacana, Los Guineos, Malay, Mabi, Mariana, Mariaco, Montegrande, Mucara, Nipe and other. For the Clay Loam texture the soil name presents are: Anones, Caguabo, Descalabrado and Morado. For the Loam texture the soils are: Coloso, Corcega, Dique, Guainabo, Mani, Maresua, Palmarejo, Reilly, Talante, Toa and other. Soils that correspond to the rock texture are: Limestone, Serpentine and Volcanic rock land, for the sand texture was found the soils: Cataño, Leveled and River wash and the last texture is Gravel which only has one soil with the same texture

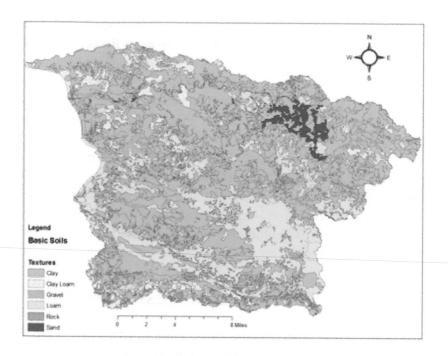

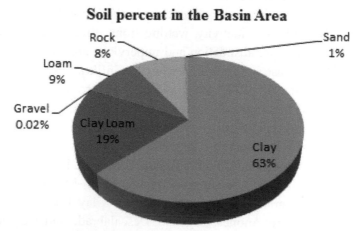

FIGURES 12.34 AND 12.35   Basic soils textures in the basin area (top: Figure 12.37), and percent of each type soil texture in the basin area (bottom: Figure 12.38).

name. Figure 12.35 presents the percentage of textures into the watershed, in which the clay encompasses most the study area with 63% of total area.

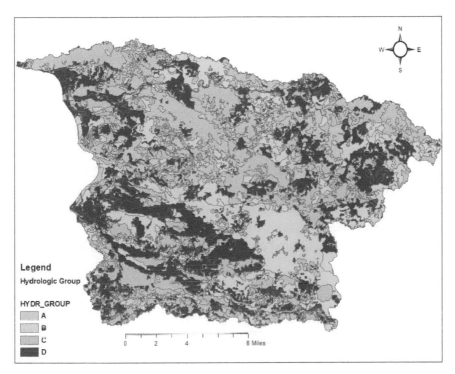

FIGURE 12.36   Hydrologic group basin map.

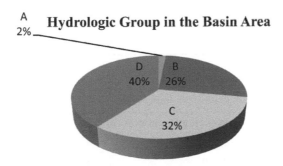

FIGURE 12.37   Percentage distribution of hydrologic group basin area.

On the other hand, the minimum texture present in the basin is the gravel with a value approximate to 0.02%.

The hydrologic group is a parameter that affects the infiltration and run-off. The Figure 12.36 presents the basin area with the hydrologic group A, B, C and D. The most representative groups are C and D, C with a 32% total area and D with a 40% total area, (Figure 12.37). These results match with the textures presented in the Figures 12.34 and 12.35, where clay and clay loam texture have more influence in the area. These soils are forming part of hydrologic groups C and D, which have the minimum infiltration rate.

Other parameters such as hydraulic conductivity, wetting front and effective porosity were assigned from literature [31, 103]. Table 12.12 presents the soils texture classification with Green-Ampt infiltration parameters. The hydraulic conductivity (K) may especially control the infiltration process when rainfall occurs over already saturated soil; the hydraulic conductivity was specified for a single layer soil profile for this study area.

The wetting front is the average capillary potential of the Green-Ampt infiltration routine, this parameter is important because it can calculate infiltration under unsaturated conditions and its value is independent of

TABLE 12.12 Green-Ampt infiltration parameter for each texture class.

Soil texture class	Effective porosity	Wetting front (cm)	Depth (cm)	Hydraulic conductivity (cm/h)
Clay	0.385	31.63	300	0.03
Clay Loam	0.309	20.88	300	0.10
Gravel	0.24	1.5	300	2.27
Loam	0.43	8.89	300	0.66
Rock	0.17	1	300	0.036
Sand	0.42	4.95	300	11.78

TABLE 12.13 Manning Roughness and Impervious

Land Use	Manning Roughness (n)	Impervious %	Area (Km²)
Agriculture	0.166	5	55.92
Agriculture /hay	0.190	4	0.12
Forest, shrub, woodland and shade coffee	0.191	2	529.12
Other emergent wetlands	0.050	1	1.24
Pasture	0.225	5	173.2
Quarries, sand and rock	0.020	95	0.56
Urban and barren	0.080	81	58.68

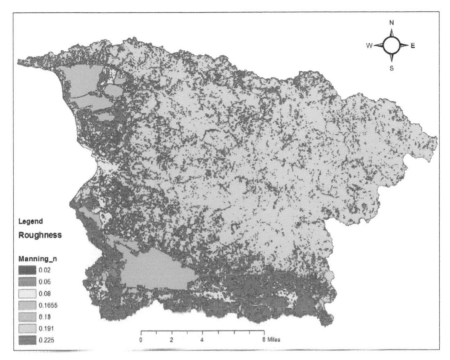

FIGURE 12.38   Map showing Manning's roughness coefficient.

soil moisture at any particular time. The effective porosity is the difference between total porosity and residual soil moisture content, this property is independent of soil moisture at any time, its range is between 1 and 0, with complete porosity being a value of one, and the value zero is for the zero porosity. The soil depth is the depth to which the infiltration can occur in the soil. If the wetting front is obstructed by a perched water table then the depth to the water table is the limiting depth. If the soil profile is limited by an impermeable layer, then the depth to that layer is the limiting depth. Soil depth can be modified through the calibration of simulations to observed stream flow [103]. The soil depth data was obtained from USDA [89–92]. Table 12.13 shows Manning's roughness coefficients. The Figure 12.38 shows Manning's roughness coefficients in the study area.

CHAPTER 13

FLOOD ALERT SYSTEM USING HIGH-RESOLUTION RADAR RAINFALL DATA: RESULTS ON DATA ACQUISITION[1, 2]

LUZ E. TORRES MOLINA

CONTENTS

13.1 INTRODUCTION

Chapters 13–19 of this book, in detail, discusses the results of the current research, which includes: data acquisition, nowcasting model results, comparison between estimation data and observation data from TropiNet radar. Hydrologic models were compared between the estimation results and observation stations data collected by the USGS. Furthermore, a comparison between rain gauges, TropiNet and NEXRAD was done. Finally the conclusions for this study are presented in Chapter 20.

---

[1] This chapter is an edited version from: *"Luz E. Torres Molina, 2014. Flood Alert System Using Rainfall Data in the Mayagüez Bay Drainage Basin, Western Puerto Rico. PhD Thesis, Department of Civil Engineering and Surveying, University of Puerto Rico, Mayagüez Campus"*.

[2] Numbers in brackets refer to the references at the end of this book.

13.2 DATA ACQUISITION

Numerous storms were analyzed during 2012 and 2014 to select the suitable storms to be forecast. Some requirements to choose the storm were defined: the data should be constantly available without interruptions, the radar should have the same elevation angle for all storms, the data may not be altered, and the radar should not stop during the storm or change its position.

All radar data storage by the system were plotted to observe the behavior and movement of the clouds. This was the first step in the selection data, graphing the data was only possible to select the data according to the

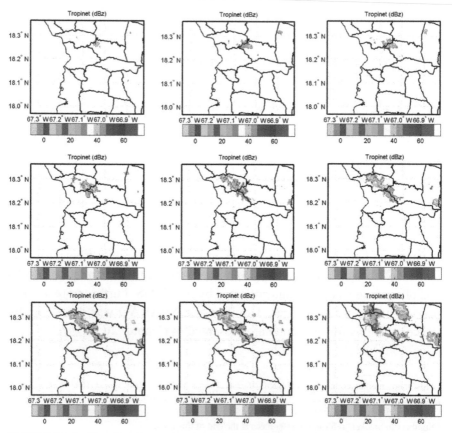

FIGURE 13.1   TropiNet storm sequence (from left to right and top to bottom).

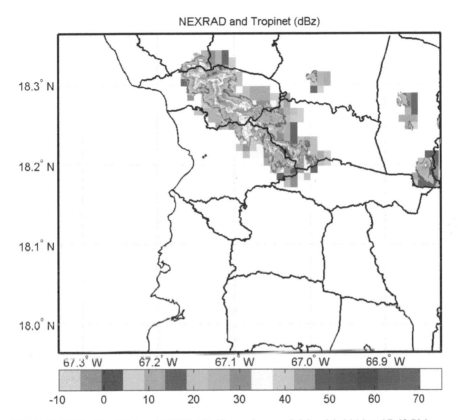

FIGURE 13.2   TropiNet and NEXRAD (Super-imposed) May 06, 2014 at 17:42 PM.

features mentioned above. In the plot, the clouds should have a time series constant with an angle of 3° for TropiNet and 0.5° for NEXRAD.

Finally, more storms were analyzed but only 10 storms were selected for this research, out of which 5 are from 2012, and the other 5 from 2014. As an example, Figure 13.1 shows TropiNet storm sequence. The Figure 13.2 shows a time series of cloud movement for the storm of May 06, 2014.

The TropiNet data was accessed from *http://weather.uprm.edu* server, and the data was raw data in binary format. Two types of transformations from binary to NetCDF were needed to handle data and from NetCDF to

Mat-file format. These transformations required the development of sub-routines in MatLab. Other transformations necessary included changing the polar coordinates to Cartesian coordinates, and it was done to handle the data in the hydrological model *Vflo* that has been discussed in Chapters 14–19.

CHAPTER 14

FLOOD ALERT SYSTEM USING HIGH-RESOLUTION RADAR RAINFALL DATA: COMPARISON AMONG RAIN GAUGES, TROPINET AND NEXRAD[1, 2]

LUZ E. TORRES MOLINA

CONTENTS

14.1   INTRODUCTION

A routine was implemented to compare the data among Rain Gauges, TropiNet and NEXRAD data. The NEXRAD pixels have 1 km^2 area and the TropiNet pixels have 60 meter for each side (0.0036 km^2 area). This means that 256 TropiNet pixels is equivalent in size to one NEXRAD pixel. In other words, within one NEXRAD pixel there are 256 TropiNet pixels. Two comparison types were done: the first was pixel to pixel, and the second was average TropiNet pixels (256) with one NEXRAD pixel.

---

[1] This chapter is an edited version from: *"Luz E. Torres Molina, 2014. Flood Alert System Using Rainfall Data in the Mayagüez Bay Drainage Basin, Western Puerto Rico. PhD Thesis, Department of Civil Engineering and Surveying, University of Puerto Rico, Mayagüez Campus"*.
[2] Numbers in brackets refer to the references at the end of this book.

14.2   COMPARISON AMONG RAIN GAUGES, TROPINET AND NEXRAD

The Figure 14.2 in Chapter 13 shows comparison and superimposed data for TropiNet and NEXRAD, for May 06, 2014 – 17:42. UPR rain gauge network is shown in Figure 14.1.

When the graphical comparison was done, the next step was to compare the rain-rate data pixels and rain gauges. Figure 14.2 (Left) presents one of many comparisons between Rain Gauge, NEXRAD and TropiNet with the original resolution at rain gauge station designate as C1 with latitude 18.2094° and longitude 67.1401°, date: May 21, 2014. The Figure 14.2 (right) also shows the comparison between NEXRAD and TropiNet at

FIGURE 14.1   UPRM Rain Gauge Network.

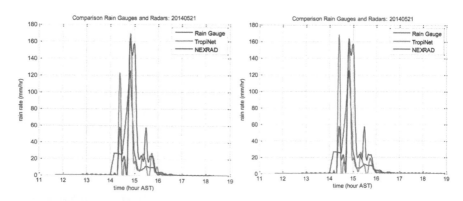

FIGURE 14.2   Left: Comparison Rain Gauge-NEXRAD and TropiNet at station C1 on, May 21, 2014 (Heavy Rain); Right: Comparison Rain Gauge-NEXRAD and TropiNet Average at station C1, on May 21, 2014 (Heavy Rain).

station C1, event May 21 of 2014, for the average pixels (256) in TropiNet, which was changed to match the resolution with NEXRAD.

As shown in Figure 14.2–14.6, the *RMS* increases under heavy rain conditions. Yet in all cases (light, moderate and heavy rain), TropiNet consistently yields the smallest error as compared to NEXRAD.

Table 14.1 includes the statistical results, where MSE is the Mean Squared Errors between Rain gauge-TropiNet and Rain gauges-NEXRAD; and RMSE is the root means squared errors. The error is greater for the comparison between rain gauge and NEXRAD data. Likewise, the best

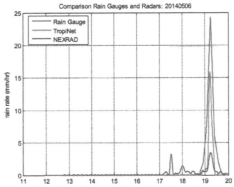

FIGURE 14.3   Comparison between Rain Gauge-NEXRAD and TropiNet at station C1, on May 06, 2014 (Moderate Rain) with original resolutions data for TropiNet and NEXRAD.

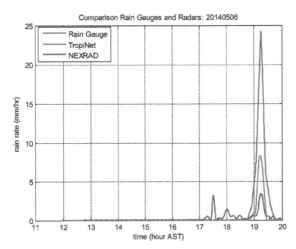

FIGURE 14.4   Comparison between Rain Gauge-NEXRAD and TropiNet Average at station C1, on May 06, 2014 (Moderate Rain) with Tropinet data degraded to match NEXRAD's

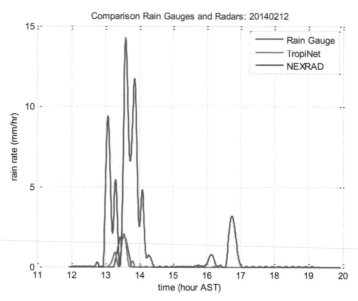

FIGURE 14.5 Comparison between rain gauge, NEXRAD and TropiNet at station C1, on February 12, 2014 (Light Rain). Values between TropiNet and rain gauge are very similar showing good agreement.

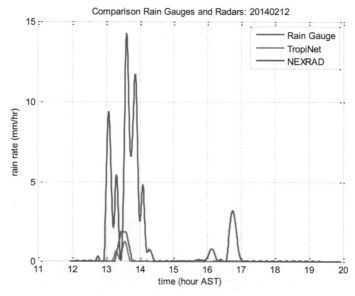

FIGURE 14.6 Comparison between rain gauge, NEXRAD and TropiNet Average at station C1, on February 12, 2014 (Light Rain). The pixel resolution for TropiNet was downgraded in order to match NEXRAD resolution. This produces larger disagreement with rain gauge data.

TABLE 14.1 Statistical Results at Station C1, on May 21, 2014

Radar-rain gauge	MSE (mm/h)^2	RMSE (mm/h)
TropiNet	344.2848	18.5549
Average TropiNet (256 pixels)	393.8165	19.8448
NEXRAD	577.1688	24.0243

result was observed between rain gauge and TropiNet data radar, when it has the original resolution (60 meters). The statistical calculations were done using following equations:

$$e_{T,i} = R_i - T_i \tag{1}$$

$$e_{N,i} = R_i - N_i \tag{2}$$

$$SSE_T = \sum_{i=1}^{n} e_{T,i}^2 \tag{3}$$

$$SSE_N = \sum_{i=1}^{n} e_{N,i}^2 \tag{4}$$

$$MSE_T = \frac{\sum_{i=1}^{n} e_{T,i}^2}{n} \tag{5}$$

$$MSE_N = \frac{\sum_{i=1}^{n} e_{N,i}^2}{n} \tag{6}$$

$$RMSE_T = \sqrt{\frac{\sum_{i=1}^{n} e_{T,i}^2}{n}} \tag{7}$$

$$RMSE_N = \sqrt{\frac{\sum_{i=1}^{n} e_{N,i}^2}{n}} \tag{8}$$

In Eqs. (1)–(8): $e_{T,i}$ and $e_{N,i}$ are the errors between Rain gauge-TropiNet and Rain gauge-NEXRAD respectively, R_i is the Rain gauge data, T_i is the TropiNet data and N_i is NEXRAD data, SSE is the sum-squared errors, the subscript T refers to TropiNet and subscript refers to NEXRAD, MSE is the mean square errors in TropiNet (T) and NEXRAD (N), and $RMSE$ is the root mean squared errors.

Figures 14.3 and 14.4 show the comparison at the station C1 but for May 06, 2014. In Figure 14.3, TropiNet with the original resolution (60 × 60 m²) presents a rain rate data with more appropriate values at C1 stations, considering rain gauge observations as the true values. This is possible due to proximity of TropiNet to the land surface and its high-resolution data. Using the simplest interpolation method, the TropiNet resolution was downgraded to the NEXRAD resolution (1 × 1 km²) (Figure 14.4). When it was compared with the other equipments, the rain rate value from TropiNet was more approximate to the NEXRAD rain rate value at the C1 station, but in more disagreement with the rain gauges. Possibly, it was due to the loss resolution.

The data tendency is very similar between TropiNet and rain gauge and NEXRAD, but NEXRAD presents significant subestimation. The statistical analysis showed that the errors were maximum when NEXRAD data was used (Table 14.2).

Other comparisons were done on February 12, 2014 at the same pixel C1. Figures 14.6 and 14.7 present precipitation distribution for rain gauge-NEXRAD-TropiNet and rain gauge-NEXRAD-TropiNet average, respectively. When the TropiNet's resolution is downgraded to match NEXRAD pixel resolution, it shows less agreement with the rain gauge values.

For this event, the tendency between TropiNet and rain gauges is the same but different to NEXRAD. The trend of TropiNet continues to be

TABLE 14.2 Statistical Results at Station C1, on May 06, 2014

Radar-rain gauge	MSE (mm/hr)²	RMSE (mm/hr)
TropiNet	4.1778	2.0439
Average TropiNet (256 pixels)	6.0680	2.4633
NEXRAD	10.8604	3.2955

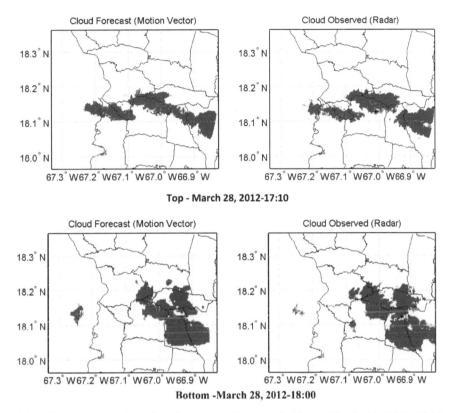

FIGURE 14.7 Cloud motion, forecast and observed: Top – March 28, 2012–17:10; bottom – on March 28, 2012–18:00.

TABLE 14.3 Statistical Results at Station C1, on February 12, 2014

Radar-rain gauge	MSE (mm/hr)²	RMSE (mm/hr)
TropiNet	0.0373	0.1931
Average TropiNet (256 pixels)	0.0428	0.2070
NEXRAD	7.0266	2.6507

more similar to rain gauges data, specifically when this radar uses its original resolutions, as shown in Table 14.3.

Other comparisons were done with different dates between 2012 and 2014. Unfortunately, 20 rain gauges were used, but only few captured good data. In most cases, rain gauges alterations to the equipment were found due to the natural or human factors.

FLOOD ALERT SYSTEM USING HIGH-RESOLUTION RADAR RAINFALL DATA: NOWCASTING MODEL MOVEMENT AND REFLECTIVITY ANALYSIS[1, 2]

LUZ E. TORRES MOLINA

CONTENTS

15.1 INTRODUCTION

In this chapter, author discusses the results of a research study on nowcasting model movement and reflectivity analysis.

There are many methods for forecasting with longer lead-time of 8, 24, and 36 h or weekly, using autoregressive methods, moving averages and others. However, the current study is a special kind of Nowcasting method for shorter lead-time in minutes. In the western Puerto Rico, sudden

[1] This chapter is an edited version from: "*Luz E. Torres Molina, 2014. Flood Alert System Using Rainfall Data in the Mayagüez Bay Drainage Basin, Western Puerto Rico. PhD Thesis, Department of Civil Engineering and Surveying, University of Puerto Rico, Mayagüez Campus*".

[2] Numbers in brackets refer to the references at the end of this book.

precipitations occur with short durations due to atmospheric conditions and topographic features at a given location. Precipitation events may develop, occur and dissipate immediately, with its duration of about 1, 2 or 3 h.

Knowing the precipitation characteristics, the nowcasting model developed in the current research only needs two lag times for prediction. This means that the model has the capacity to forecast the rainfall even if the duration is very short. The developed model is presenting the best prediction when the lead-time is 10 min. The postulated rainfall nowcasting algorithm involves two major tasks: (a) predicting the future location of the rain pixels, and (b) predicting rainfall at each pixel.

15.2   NOWCASTING MODEL MOVEMENT AND REFLECTIVITY ANALYSIS

Figure 7.7 (top) in Chapter 7 shows the cloud motion comparison between observed (right) movement and estimated (left) movement at storm date March 28, 2012, 17:10 h. The black point is the centroid at initial time and the red point is the centroid at the final time. In some cases there is more than one cloud centroid, and therefore there is more than one black and red point in this Figure. This happens when the division cloud method has detected more than one cloud system within the area. Figure 7.7 (bottom) presents the separation cloud with two centroids at cloud forecast, storm date March 28, 2012 18:00 h.

In this chapter, Figure 15.1 presents the sequence of event during 40 min considering each ten min of cloud motion within a total duration event of 7 h where t_o = 16:50 h, on March 28, 2012. Figure 15.2 has the same sequence with a lead-time of 20 min where t_o = 17:10 hr. In this case the sequence of event during 80 min was considered. Figure 15.3 shows 120 min of the event, the sequence for a lead-time of 30 min where t_o = 17:30 hr. Other storms were processed in the same way. The figures and results are in the office "Red de Radares del Tiempo" University of Puerto Rico at Mayagüez.

The comparison of estimated or predicting reflectivity using the main Eq. (5) and observer reflectivity at each pixel were furthermore performed. Figure 15.4 shows the comparison with a lead-time of 10 min where t_o = 16:50 hr.

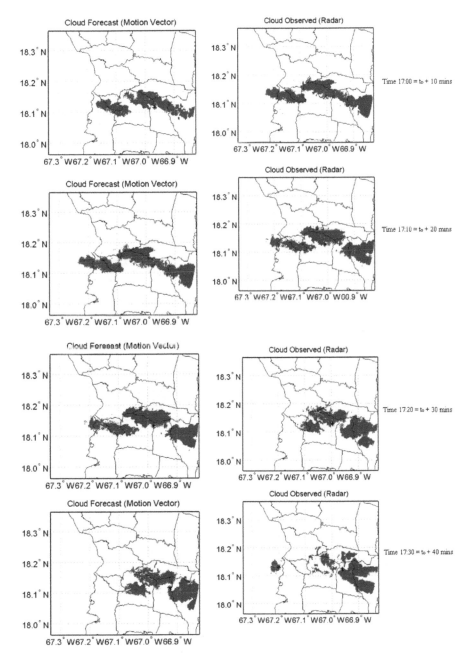

FIGURE 15.1 Cloud motion sequence with a lead-time of 10 min, $t_o = 16{:}50$ hr, on March 28, 2012.

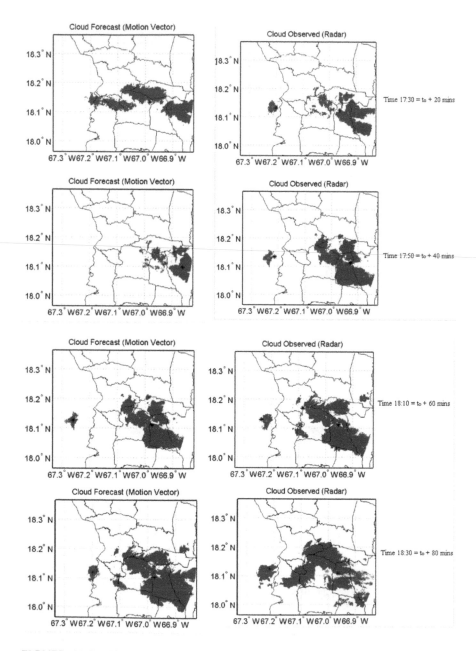

FIGURE 15.2 Cloud motion sequence with a lead-time of 20 min, $t_o = 17{:}10$ hr, on March 28, 2012.

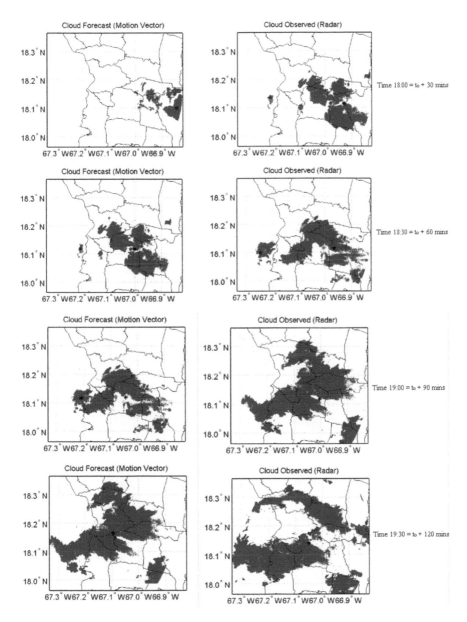

FIGURE 15.3 Cloud motion sequence with a lead-time of 30 min, $t_o = 17:30$ hr, on March 28, 2012.

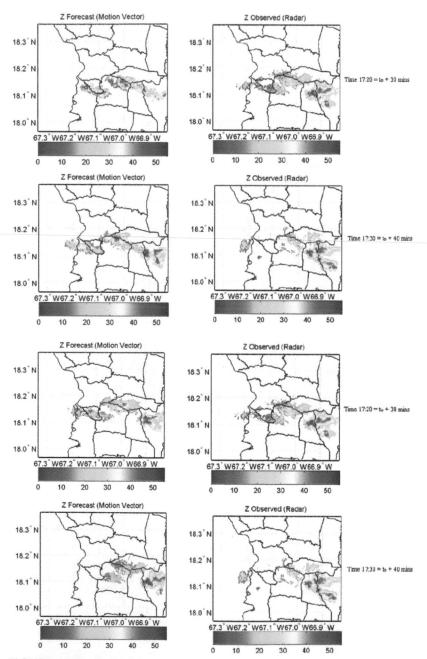

FIGURE 15.4 Reflectivity sequence with a lead-time of 10 min, t_o = 16:50 hr on March 28, 2012.

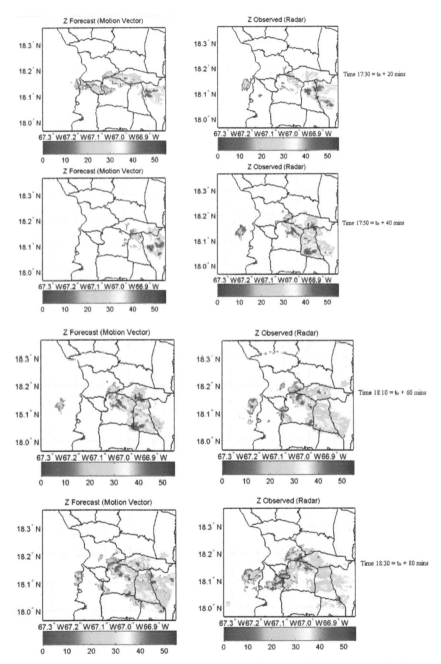

FIGURE 15.5   Reflectivity sequence with a lead-time of 20 min, t_o = 17:10 hr on March 28, 2012.

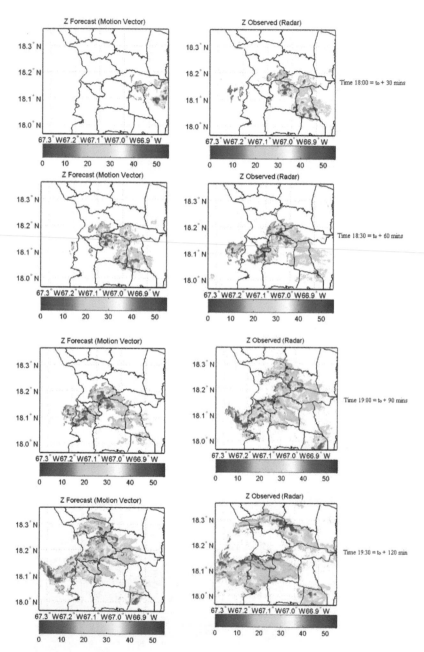

FIGURE 15.6 Reflectivity sequence with a lead-time of 30 min, t_o = 17:30 hr on March 28, 2012.

Figure 15.5 presents a comparison between estimated and observed data but with a lead-time of 20 min of the event where $t_o = 17{:}10$ hr. Finally, Figure 15.6 is with a 30 min lead-time where $t_o = 17{:}30$ hr.

For all events, the best results were presented with a prediction of 10 min (Figure 15.4). Western Puerto Rico area geographical position makes it susceptible to sudden rainfalls that are changing rapidly in time and space. Due to this change, a lead-time of 10 min is the time prediction more adequate to this precipitation class. A larger lead-time results in greater statistical errors. Contrarily using a lead-time smaller than 10 min, the purpose of flood alert system will be annulled by the absence of time to evacuation.

It is important to mention that the algorithm to forecast precipitation uses a sequence of the observed rainfall data to estimate the movement direction and size of the cloud or cell. And then using the main Eq. (5) in Chapter 12, rainfall is estimated in each pixel within every zone. Thereby, the suggested regression model was developed under the following assumption. It is expected that in a short time (10 min) period a rain cloud behaves approximately as a rigid object and the cloud rain pixels moves in a constant speed and direction. Thus, the most likely future rainfall areas can be estimated by using the advection of the centroids of the rain cells in consecutive images. The current estimation reflectivity is a function of the previous reflectivity images observed. Rainfall nowcasting algorithm task is predicting rainfall rate at each pixel.

CHAPTER 16

FLOOD ALERT SYSTEM USING HIGH-RESOLUTION RADAR RAINFALL DATA: ESTIMATION OF PARAMETERS[1, 2]

LUZ E. TORRES MOLINA

CONTENTS

16.1   INTRODUCTION

This methodology was applied to estimate four unknown parameters ($\delta1$, $\delta2$, $\delta3$, and Φ) so as to:

- Find the optimum values with a bounded constraint: first linearized the main equation;
- Identify the initial point trough a nonlinear regression model where the phi Φ is temporarily ignored, and the deltas values initial are obtained by solving the linear regression; and

---

[1] This chapter is an edited version from: "*Luz E. Torres Molina, 2014. Flood Alert System Using Rainfall Data in the Mayagüez Bay Drainage Basin, Western Puerto Rico. PhD Thesis, Department of Civil Engineering and Surveying, University of Puerto Rico, Mayagüez Campus*".

[2] Numbers in brackets refer to the references at the end of this book.

• Find the optimum values using a constrained nonlinear optimization technique to estimate the final parameter set for each zone (9 × 9) and every window where the phi Φ parameter is a bias correction factor introduced in the optimization.

16.2   ESTIMATION OF PARAMETERS

The optimum parameters for the nonlinear regression model were estimated by solving a constrained nonlinear optimization problem (*fmincon*), as shown in Table 16.1.

The derived initial point was ingested into the constrained nonlinear subroutine to facilitate convergence, the delta parameters were restricted to be positives and phi parameter was restricted to be in the range of 0 to 1.1 values. For purposes of demonstration, Table 16.1 presents the initial point and final point of the estimated parameters ($\delta 1$, $\delta 2$, $\delta 3$, and Φ) for a random zone (9 × 9) that occurred on March 28, 2012.

Figure 16.1 shows the distribution of initial and optimal values of phi (Φ) with a lead-time of 10 min. For the comparison between the parameters, initial deltas and optimal deltas were used as a statistic test (T-statistics) to determine whether or not the optimization causes a change in mean values. If the optimum mean values are significantly different from the original mean values, it is possible to conclude that the treatment has a significant effect. Figure 16.2 presents the median phi coefficient for the initial value and optimal value.

TABLE 16.1   Parameter Estimation for a Random Zone (9 × 9), on March 28, 2012 for a Lead-Time of 10 min

Parameter	Initial point (Linear Regression)		Nonlinear regression
	Estimation	T-statistics	Final Estimation
$\delta_{1,k}$	0.03546	0.65098	0.00507
$\delta_{2,k}$	0.06596	2.89453	0.47448
$\delta_{3,k}$	−2.47237	−1.01741	0.00012
φ_k	2.18039	–	0.81903
RMSE$_t$	29.51233	–	2.01960

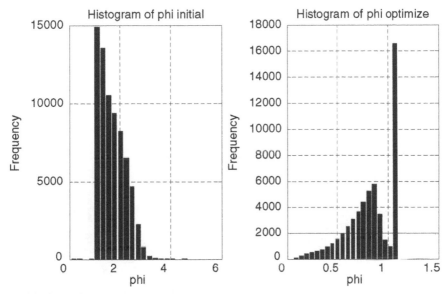

FIGURE 16.1 Distribution of initial value of phi (left) and the optimal values of phi (right) for the storm date: March 28, 2012, for a lead-time of 10 min.

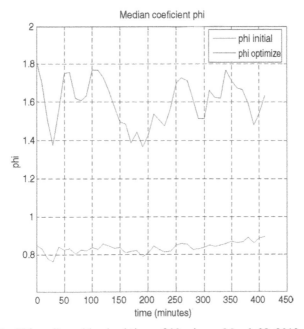

FIGURE 16.2 Phi median with a lead-time of 10 min, on March 28, 2012.

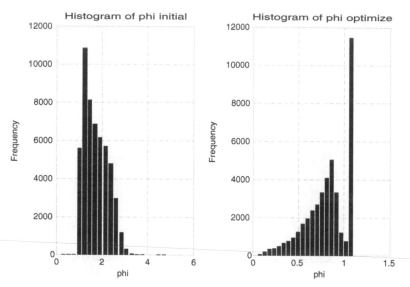

FIGURE 16.3 Distribution of initial value of phi (left) and the optimal values of phi (right) for the storm date: March 28, 2012, for a lead-time of 20 min.

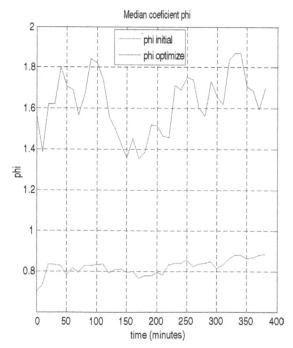

FIGURE 16.4 Phi median with a lead-time of 20 min, on March 28, 2012.

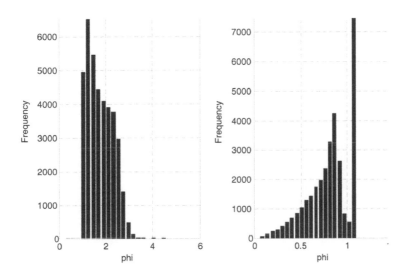

FIGURE 16.5 Distribution of initial value of phi (left) and the optimal values of phi (right) for the storm date: March 28, 2012, for a lead-time of 30 min.

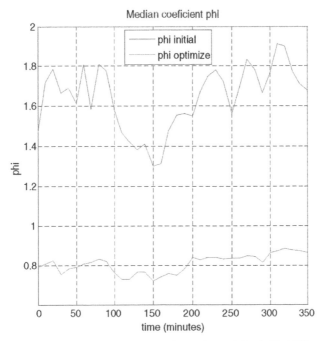

FIGURE 16.6 Phi median with a lead-time of 30 min, on March 28, 2012.

Figure 16.3 presents the distribution of initial variable phi (Φ) and the optimal value for a lead-time of 20 min and Figure 16.4 is the median of the value phi for lead-time 20 min. Figure 16.5 presents the distribution of initial variable phi (Φ) and the optimal value for a lead-time of 30 min. And Figure 16.6 is the median of the value phi for lead-time 30 min. The analysis was made for all storms (10) and similar results were obtained.

FLOOD ALERT SYSTEM USING HIGH-RESOLUTION RADAR RAINFALL DATA: NOWCASTING MODEL VALIDATION[1, 2]

LUZ E. TORRES MOLINA

CONTENTS

17.1 INTRODUCTION

An analysis for the nowcasting requires a combination of meteorological and hydrological statistics, as this permits a better understanding of behavior of the spatial and temporal accuracy of storm prediction. A good

---

[1] This chapter is an edited version from: *"Luz E. Torres Molina, 2014. Flood Alert System Using Rainfall Data in the Mayagüez Bay Drainage Basin, Western Puerto Rico. PhD Thesis, Department of Civil Engineering and Surveying, University of Puerto Rico, Mayagüez Campus"*.

[2] Numbers in brackets refer to the references at the end of this book.

nowcasting include accuracy of the spatial, as well as in the temporal level and accuracy of the predicted rainfall intensity. Model performance criteria for the prediction required quantitative comparison measures, these measures include ten storms mentioned before in Table 12.6 in Chapter 12.

17.2   NOWCASTING MODEL VALIDATION

The accuracy of rainfall prediction of each pixel can be measured by decomposing the rainfall process into sequences of discrete and continuous random variables, i.e., the presence or absence of rainfall events and rainfall intensity. Examples of quantitative parameters used in the current research include: Contingency table, Mean square Error (MSE), Root Mean Square Error (RMSE), Bias Ratio (BR) and Mean Absolute Error (MAE). These parameters will be discussed in detail below.

The joint distribution of the forecast and observations has fundamental interest with respect to the verification of forecasts. In the most practical setting, both the forecast and observations are discrete variables. Even if the forecasts and observations are not already discrete quantities. Denote the forecast by y_i, which can take on any of the possible I values (y_1, y_2, y_3,..., y_I); and the corresponding observations as O_j, which can take on any of the possible J values (O_1, O_2, O_3,..., O_J). Then the joint distribution of the forecast and observation is denoted as:

$$p\left(y_i, O_j\right) = \Pr\left\{y_i, O_j\right\} = \Pr\left\{y_i \cap O_j\right\}; i = 1,...I; j = 1,...J \qquad (1)$$

This is a discrete bivariate probability distribution function, associating a probability with each of the $I \times J$ possible combinations of forecast and observation [104]. The contingency table (Figure 17.1) $I \times J$ shows the arrangement of four possible combinations of forecast/event pairs for a simple $I = J = 2$ case.

17.3   ATTRIBUTES RELATED WITH THE CONTINGENCY TABLE

Hit rate (HR) is the ratio of correct forecasts to the number of times this event can occur, as shown in the following equation:

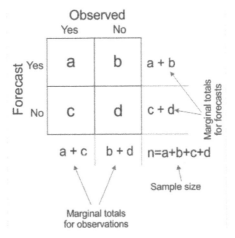

FIGURE 17.1   Contingency table based on *Wilks* [104].

$$HR = \frac{(a+d)}{(a+b+c+d)} \qquad (2)$$

The probability of detection (*POD*) as the fraction of those occasions when the forecast event occurred on which it was furthermore forecasted, in this case it is the probability that rain occur.

$$POD = \frac{a}{a+c} \qquad (3)$$

The False Alarm Ratio (*FAR*) is the relation of the forecast events that fail to materialize: the best possible *FAR* is zero and the worst possible *FAR* is one.

$$FAR = \frac{b}{a+b} \qquad (4)$$

$$Bias = \frac{(a+b)}{(a+c)} \qquad (5)$$

The Bias (*Bias*) is the ratio of the number of yes forecasts to the number of yes observed. Unbiased forecast exhibit *Bias* = 1, indicating that the event forecasted the same number of times that it was observed [104].

Table 17.1 shows the contingency table for the storm of March 28, 2012 with a lead-time of 10 min, 20 min and 30 min. Table 17.2 shows the contingency table for the storm of March 29, 2012 with a lead-time of 10 min, 20 min and 30 min.

Table 17.3 shows the contingency table for the storm of April 30, 2012 with a lead-time of 10 min, 20 min and 30 min. Table 17.4 shows the contingency table for the storm of October 10, 2012 with a lead-time of 10 min, 20 min and 30 min. Table 17.5 shows the contingency table for the storm of February 12, 2014 with a lead-time of 10 min, 20 min and 30 min. Table 17.6 shows the contingency table for the storm of May 06, 2014 with a lead-time of 10 min, 20 min and 30 min. Table 17.7 shows the contingency table for the storm of May 21, 2014 with a lead-time of 10 min, 20 min and 30 min.

Table 17.8 shows the contingency table for the storm of June 29, 2014 with a lead-time of 10 min, 20 min and 30 min. Table 17.9 shows the contingency table for the storm of June 30, 2014 with a lead-time of 10 min, 20 min and 30 min. Table 17.10 shows the contingency table for the storm of July 05, 2014 with a lead-time of 10 min, 20 min and 30 min. Finally, Table 17.11 shows the average the contingency table associated with the ten studied storms.

The performance index is introduced in this research to measure the overall dichotomous (rain/no rain) forecast accuracy of the model, and is computed as a function of *HR, FAR* and *POD*. The performance index varies from zero to one, and a value of one correspond to the best algorithm performance; whereas, zero corresponds to the worst case. The performance index (*PI*) is defined as follows:

$$PI = 1 - \frac{FAR - POD - HR + 2}{3} \qquad (6)$$

Table 17.12–17.14 present model performance score: *HR, POD, FAR,* Detection Bias (*DB*) and PI for the ten storms with 10 min, 20 min and 30 min of lead-time. And finally Table 17.15 shows the average of detection results for all storms to the model or the performance score of all storms.

TABLE 17.1 Contingency Table for the Storm that Occurred on March 28, 2012

		Contingency Table					
		10 min		20 min		30 min	
		Observed		Observed		Observed	
Forecast		Yes	No	Yes	No	Yes	No
	Yes	5382890	1404061	2285726	896220	1197386	599598
	No	2173265	20221270	1473238	9241282	1125112	5415278

TABLE 17.2 Contingency Table for the Storm that Occurred on March 29, 2012

		Contingency Table					
		10 min		20 min		30 min	
		Observed		Observed		Observed	
Forecast		Yes	No	Yes	No	Yes	No
	Yes	807073	406387	220028	210810	104527	136244
	No	479654	22646027	249631	10445985	153440	6559901

TABLE 17.3 Contingency Table for the Storm that Occurred on April 30, 2012

Contingency Table

Forecast	10 min		20 min		30 min	
	Observed		Observed		Observed	
	Yes	No	Yes	No	Yes	No
Yes	1455552	607776	465605	357759	170531	218996
No	821678	13769757	432562	6382857	250354	3520858

TABLE 17.4 Contingency Table for the Storm that Occurred on October 10, 2012

Contingency Table

Forecast	10 min		20 min		30 min	
	Observed		Observed		Observed	
	Yes	No	Yes	No	Yes	No
Yes	4285411	1714603	1733340	996010	1733340	996010
No	2440495	13066736	1499016	5481068	1499016	5481068

TABLE 17.5 Contingency Table for the Storm that Occurred on February 12, 2014

Contingency Table

		10 min			20 min			30 min	
		Observed			Observed			Observed	
Forecast		Yes	No	Forecast	Yes	No	Forecast	Yes	No
	Yes	2219965	1143951	Yes	797385	721777	Yes	375944	486929
	No	1976191	25961893	No	1252900	11835538	No	856713	6627614

TABLE 17.6 Contingency Table for the Storm that Occurred on May 06, 2014

Contingency Table

		10 min			20 min			30 min	
		Observed			Observed			Observed	
Forecast		Yes	No	Forecast	Yes	No	Forecast	Yes	No
	Yes	899183	582657	Yes	243862	374693	Yes	125040	238542
	No	1126297	25214088	No	734163	11862788	No	500325	7482572

TABLE 17.7 Contingency Table for the Storm that Occurred on May 21, 2014

		Contingency Table				
Forecast		**10 min**		**20 min**		
		Observed		**Observed**		
		Yes	No		Yes	No
	Yes	1589477	733786	**Forecast** Yes	608566	436933
	No	1295752	20029060	No	780853	9301986

		30 min	
		Observed	
Forecast		Yes	No
	Yes	335552	324178
	No	558082	5041851

TABLE 17.8 Contingency Table for the Storm that Occurred on June 29, 2014

		Contingency Table				
Forecast		**10 min**		**20 min**		
		Observed		**Observed**		
		Yes	No		Yes	No
	Yes	567993	257798	**Forecast** Yes	199172	173581
	No	348501	10650825	No	237983	4258457

		30 min	
		Observed	
Forecast		Yes	No
	Yes	70192	138215
	No	150575	2423418

TABLE 17.9 Contingency Table for the Storm that Occurred on June 30, 2014

		Contingency Table						
		10 min			20 min		30 min	
		Observed			Observed		Observed	
Forecast		Yes	No	Forecast	Yes	No	Yes	No
	Yes	2427073	656663	Yes	979092	477523	471969	311968
	No	1426931	11482242	No	826194	4469956	577619	2810500

TABLE 17.10 Contingency Table for the Storm that Occurred on July 05, 2014

		Contingency Table						
		10 min			20 min		30 min	
		Observed			Observed		Observed	
Forecast		Yes	No	Forecast	Yes	No	Yes	No
	Yes	1694067	622202	Yes	632278	437548	339874	320228
	No	1096379	7714503	No	626156	3171949	371393	1750199

TABLE 17.11 Contingency Table for 10 Storms

Contingency Table

		10 min		20 min		30 min	
		Observed		Observed		Observed	
Forecast		Yes	No	Yes	No	Yes	No
Forecast	Yes	21328684	8129884	8165054	5082854	4198082	3601592
	No	13185143	170756401	8112696	76651866	5806632	48954746

TABLE 17.12 Detection Results for 10 Storms with a Lead-Time of 10 min

Skill Score					Forecast					
Dates	20120328	20120329	20120430	20121010	20140212	20140506	20140521	20140629	20140630	20140705
HR	0.87741	0.96359	0.91417	0.80680	0.90032	0.93857	0.91417	0.94872	0.86971	0.84555
POD	0.71238	0.62722	0.63917	0.63715	0.52904	0.44393	0.55090	0.61974	0.62975	0.60709
FAR	0.20687	0.33489	0.29456	0.28576	0.34006	0.39319	0.31584	0.31218	0.21294	0.26862
Detection Bias	0.89820	0.94305	0.90606	0.89207	0.80166	0.73159	0.80522	0.90103	0.80013	0.83007
PI	0.79430	0.75197	0.75292	0.71939	0.69643	0.66310	0.71641	0.75209	0.76217	0.72800

TABLE 17.13 Detection Results for 10 Storms with a Lead-Time of 20 min

Skill Score	Forecast									
Dates	20120328	20120329	20120430	20121010	20140212	20140506	20140521	20140629	20140630	20140705
HR	0.82949	0.95861	0.89653	0.74303	0.86481	0.91609	0.89056	0.91547	0.81248	0.78148
POD	0.60807	0.46848	0.51839	0.53624	0.38891	0.24934	0.43800	0.45560	0.54234	0.50243
FAR	0.28165	0.48930	0.43450	0.36492	0.47511	0.60575	0.41791	0.46567	0.32783	0.40898
Detection Bias	0.84649	0.91734	0.91671	0.84438	0.74095	0.63245	0.75247	0.85267	0.80686	0.85012
PI	0.71863	0.64593	0.66014	0.63811	0.59287	0.51989	0.63688	0.63513	0.67566	0.62497

TABLE 17.14 Detection Results for 10 Storms with a Lead-Time of 30 min

Skill Score	Forecast									
Dates	20120328	20120329	20120430	20121010	20140212	20140506	20140521	20140629	20140630	20140705
HR	0.79313	0.95834	0.88735	0.69252	0.83903	0.91147	0.85905	0.89620	0.78677	0.75136
POD	0.51555	0.40519	0.40517	0.43313	0.30498	0.19994	0.37549	0.31794	0.44967	0.47789
FAR	0.33366	0.56586	0.56221	0.45758	0.56431	0.65608	0.49137	0.66319	0.39795	0.48510
Detection Bias	0.77372	0.93334	0.92549	0.79853	0.70001	0.58139	0.73825	0.94401	0.74689	0.92813
PI	0.65834	0.59922	0.57677	0.55602	0.52656	0.48511	0.58105	0.51698	0.61283	0.58139

TABLE 17.15   Detections Results: Model Accuracy Score Considering All Events
as a Single Group

Skill Score	Forecast		
	10 min	**20 min**	**30 min**
HR	0.90011	0.86536	0.84961
POD	0.61797	0.50160	0.41961
FAR	0.27597	0.38367	0.46176
Bias	0.85352	0.81386	0.77959
PI	0.74737	0.66110	0.60248

17.4   HIT RATE

For lead-times of 10, 20 and 30 min, the storms provide an average hit rate
(HR) of 0.90, 0.86 and 0.84, respectively. The hit rate score is the fraction
of observed events that is forecast correctly. It ranges from zero (0) at the
poor end to one (1) at the good end. The probability of detection (POD)
of storms varies from 0.61, 0.50 and 0.41. While the false alarm rates
(FAR) is 0.27, 0.38 and 0.46 for lead-time of 10, 20 and 30 min, respectively. Figure 17.2 shows POD values and FAR values for the complete

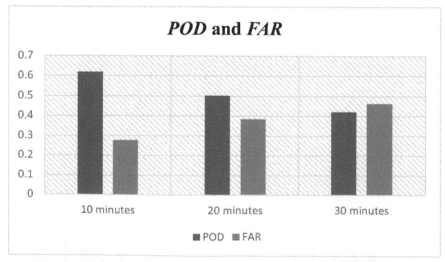

FIGURE 17.2   Probability of detection and false alarm for the all storms.

set of storms. In the ideal situation *POD* should approach to one (1), while the *FAR* results should approach to zero (0). The performance index was 0.74, 0.66 and 0.60 for 10 min, 20 min and 30 min, respectively for the model, (Figure 17.3).

Similarly, the Hit Rate (*HR*) of the model for the all storms was 0.90, 0.86 and 0.84 for the 10, 20 and 30 min, respectively (Figure 17.4). Other strategy for validations was made: In this case the validation is for the quantity of rainfall estimation, by comparing each pixel predicted of rainfall intensity at a given time and a given specific lead-time with the corresponding observed rainfall intensity.

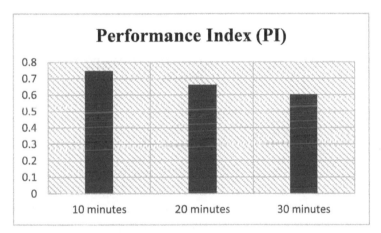

FIGURE 17.3 Performance Index for the all storms.

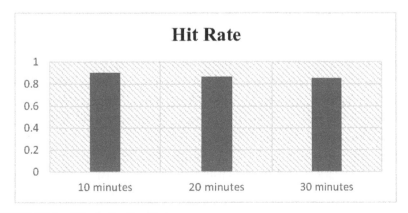

FIGURE 17.4 Hit rate for the all storms.

17.4.1 ROOT MEAN SQUARE ERROR (RMSE) AND BIAS RATIO

These results were analyzed using the Root Mean Square Error (*RMSE*) and Bias Ratio (*BR*), as a mean for the estimation quantity. The calculation of these scores is given as follows:

$$RMSE_{t+l} = \sqrt{\frac{\sum_{i=1}^{n*}\sum_{j=1}^{m}\left[y_{t+l}(i,j) - \hat{y}_{t+l}(i,j)\right]^2}{n*m}}, for\, l = 10, 20, 30 \quad (7)$$

$$\overline{RMSE_l} = \frac{\sum_{t=1}^{N} RMSE_{t+l}}{N}, for\, l = 10, 20, 30 \quad (8)$$

$$BR_{t+l} = \frac{\sum_{i=1}^{n*}\sum_{j=1}^{m}\hat{y}_{t+l}(i,j)}{\sum_{i=1}^{n*}\sum_{j=1}^{m}y_{t+l}(i,j)}, for\, l = 10, 20, 30 \quad (9)$$

$$\overline{BR_l} = \frac{\sum_{t=1}^{N} BR_{t+l}}{N}, for\, l = 10, 20, 30 \quad (10)$$

where, $\hat{y}_{t+l}(i,j)$ is the predicted rainfall intensity made at time t with lead-time l units for a pixel located at (i,j), and $y_{t+1}(i,j)$ is the corresponding observed rainfall intensity, N is the total number of units of time that rainfall was observed, $n*$ is the total number of rows and m is total number of columns of rainfall area.

The *RMSE* and *BR* for each event with a lead-time of 10, 20 and 30 min are given in Tables 17.16–17.18.

The Root mean square error (*RMSE*) and Bias ratio (*BR*) measure the accurate of the simulation for all ten studied events are given in Table 17.19, which furthermore shows the corresponding average values for each lead-time 10, 20 and 30 min, respectively. The *RMSE* average values are 0.026, 0.077 and 0.144 mm and the Bias average values are 0.97, 0.98 and 1.04 for lead-times of 10, 20 and 30 min, respectively.

The estimation Bias ratio for a lead-time of 30 min presents an average over estimation prediction, while the estimation Bias ratio for a lead-time of 10 min and 20 min shows sub estimation. The Bias ratio for the three lead-times is near to one; this means that they are good estimates [60].

TABLE 17.16 Estimation Results with a Lead-Time of 10 min

	Estimation Results– Lead-time 10 min									
	Forecast									
Event	20120328	20120329	20120430	20121010	20140212	20140506	20140521	20140629	20140630	20140705
RMSE (mm)	0.04141	0.01014	0.00799	0.05705	0.02575	0.01287	0.01899	0.03161	0.03220	0.02925
Estimation Bias	0.94546	0.97275	1.02718	0.95023	0.96658	0.89165	0.92123	1.10117	0.90473	1.04828

TABLE 17.17 Estimation Results with a Lead-Time of 20 min

	Estimation Results– Lead-time 20 min									
	Forecast									
Event	20120328	20120329	20120430	20121010	20140212	20140506	20140521	20140629	20140630	20140705
RMSE (mm)	0.13834	0.03145	0.03299	0.14595	0.06009	0.02856	0.05958	0.10140	0.07244	0.10681
Estimation Bias	0.91111	0.97859	1.11749	0.92434	0.91703	0.82576	0.90065	1.04347	0.99950	1.21564

TABLE 17.18 Estimation Results with a Lead-Time of 30 min

	Estimation Results– Lead-time 30 min									
	Forecast									
Event	20120328	20120329	20120430	20121010	20140212	20140506	20140521	20140629	20140630	20140705
RMSE (mm)	0.20653	0.03478	0.04309	0.36443	0.15250	0.04805	0.11523	0.15488	0.16444	0.16347
Estimation Bias	0.82901	0.97343	1.19152	0.85672	0.84179	0.78234	0.83742	1.37135	0.98537	1.80007

TABLE 17.19   Average Root Mean Square Error and Bias Rate for 10 Events

Estimation Results			
Forecast Errors Average			
Lead-time	10 min	20 min	30 min
RMSE (mm)	0.02673	0.07776	0.14474
Estimation Bias Ratio	0.97293	0.98336	1.04690

The *RMSE* average in 10 min lead-time presents the best result compared with the other lead-time of 20 min and 30 min. The *RMSE* is increasing due to the fact that large errors are occurring because the lead-time is increasing.

17.5 ACCUMULATION OF RAINFALL

Figure 17.5 shows the accumulation of rainfall for the first five events with a lead-time of 10 min, and Figure 17.6 presents the last five events. This accumulation is for every pixel and total duration. The storm duration is different for each date. The left panel shows the accumulated predicted rainfall in millimeters for 10 min of lead-time and the right panel shows TropiNet observed accumulated rainfall with a lead- time of 10 min.

Figures 17.7 and 17.8 show the average rainfall for all rain pixels during each time interval (10 min) for the events. In these figures, it is possible to observe a time shift due to cloud velocity movement. In this methodology the velocity was assumed as constant for each event. Corfidi et al. [11] determined that velocity in the convective systems required two components: the cell velocity of the system and the propagation velocity due to occurrence, development and merger of the convective cell. The most difficult task is determined the propagation velocity [11]. The time shift deficiency due the absence of atmospherics factors to evaluate the propagation velocity, could be fixed with mean time shift estimation for all storms depending of their lead-time.

The forecast results present the same tendency of the observed data where the peaks with more precipitation in TropiNet events are coinciding with the forecasted data. They are in good agreement considering that the prediction is in short time and space.

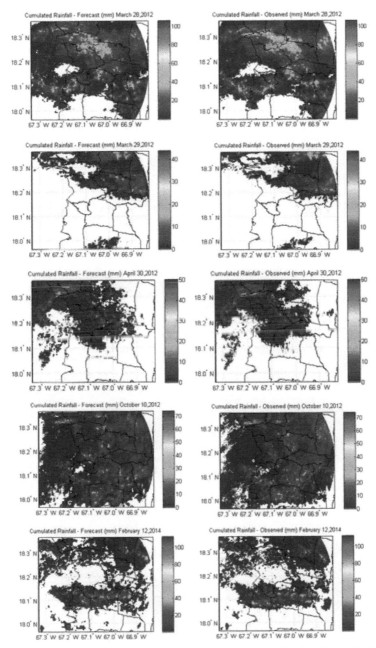

FIGURE 17.5 Rainfall accumulated during the each event, the first 5 events. The left column is the forecasted cumulated rainfall and the right column is the observed cumulated rainfall.

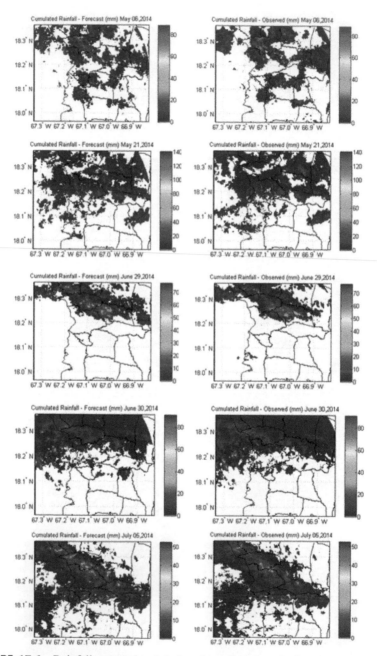

FIGURE 17.6   Rainfall accumulated during the each event, the last 5 events. The left column is the forecasted cumulated rainfall and the right column is the observed cumulated rainfall.

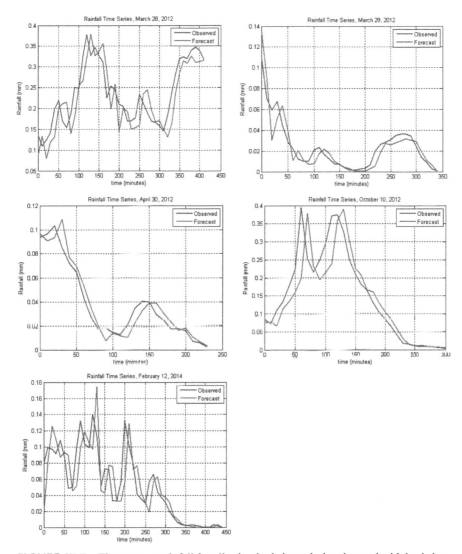

FIGURE 17.7 The average rainfall for all rain pixels in each time interval with lead-time of 10 min during the first 5 events. The blue line represents the observed data (TropiNet) and the green line represents the forecasted data accumulated precipitation for all rain pixel along the total storm event.

Figure 17.9 present for the first five events, left panel is the accumulated average rainfall for all rain pixels during the total event. It was calculated

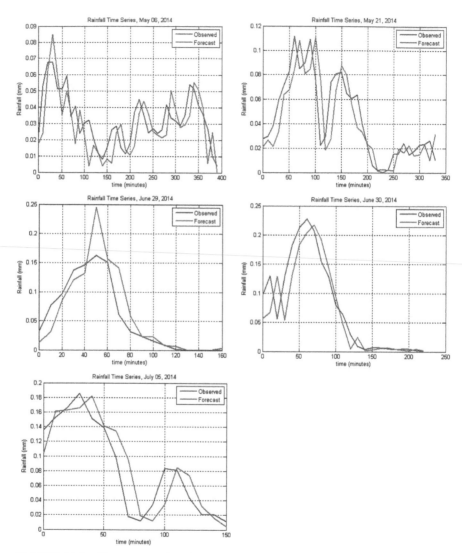

FIGURE 17.8   The average rainfall for all rain pixels in each time interval with lead-time of 10 min during the last 5 events. The blue line represents the observed data (TropiNet) and the green line represents the forecasted data accumulated precipitation for all rain pixel along the total storm event.

taking the rainfall total during the storm and the precipitation total area. The right panel is the scatter plot at the same rainfall event. Similarly,

Figure 17.10 shows the results for last five events. These figures show that model exhibits a small underestimation in all events. But it is possible to perceive in general that the forecast is highly similar to the observed data. They have the same tendency in the time series during all events.

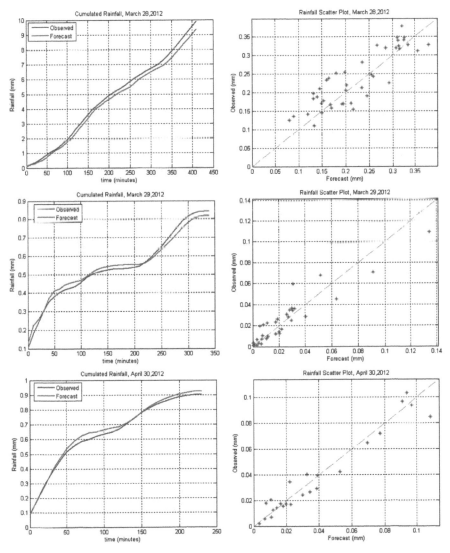

FIGURE 17.9 Continued

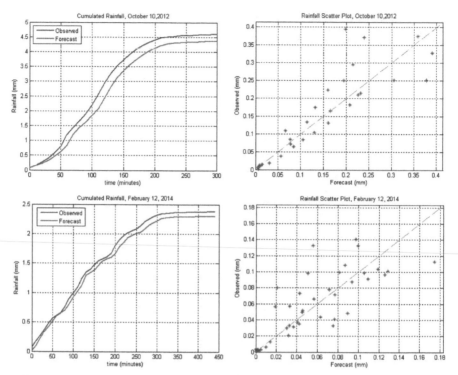

FIGURE 17.9 Left panel shows the accumulated precipitation average for all rain pixels during the all rainfall events. The blue line represents the observed precipitation and the green line the forecast. The right panel shows the corresponding scatter plot of the same rainfall event (first 5 events).

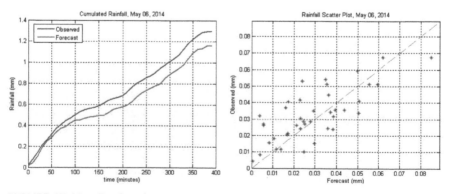

FIGURE 17.10 Continued

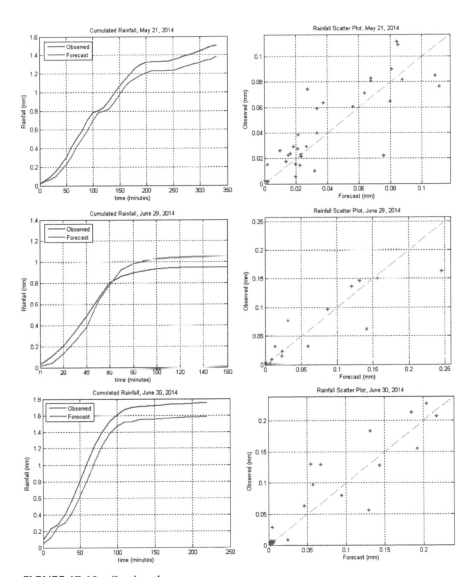

FIGURE 17.10 Continued

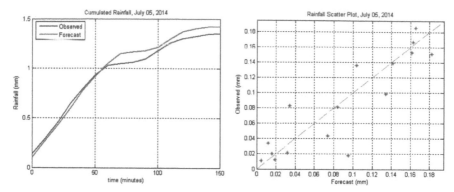

FIGURE 17.10   Left panel shows the accumulated precipitation average for all rain pixels during the total events. The blue line represents the observed precipitation and the green line the forecast. The right panel shows the corresponding scatter plot of the same rainfall event (last 5 events).

17.6   SUMMARY

The rainfall nowcasting algorithm uses consecutives images of weather radar to forecast rainfall rate. The algorithm searches for contiguous rain pixels and identifies rain cells in the last two radar images to estimate the cloud motion vector. The cloud motion vector is then used to estimate the most likely future locations of the rain pixels, and finally, nonlinear regression models are developed to forecast the intensity of rainfall rate at each rain pixel. The new rainfall nowcasting algorithm was validated with ten storms and results show that the nowcasting algorithm is a potential tool to couple with a hydrological numerical model to predict the most likely inundation areas.

CHAPTER 18

FLOOD ALERT SYSTEM USING HIGH-RESOLUTION RADAR RAINFALL DATA: VALIDATION OF HYDROLOGIC MODEL VFLO[1, 2]

LUZ E. TORRES MOLINA

CONTENTS

18.1 INTRODUCTION

The hydrological model *Vflo* required the ensemble of various layers that perform the physical and topographic characteristics of the basin area. These layers are formed by parameters that were previously presented as: effective porosity, hydraulic conductivity, wetting front, roughness, soil depth, and initial saturation which can be most sensitive in the watershed. Spatially distributed parameter and input from radar rainfall requires new methods for adjustment in order to minimize differences between simulated and observed hydrographs. The hydraulic roughness (n), hydraulic conductivity

---

[1] This chapter is an edited version from: "*Luz E. Torres Molina, 2014. Flood Alert System Using Rainfall Data in the Mayagüez Bay Drainage Basin, Western Puerto Rico. PhD Thesis, Department of Civil Engineering and Surveying, University of Puerto Rico, Mayagüez Campus*".

[2] Numbers in brackets refer to the references at the end of this book.

(K) and initial saturation (θ) are the most sensitive parameters of the hydrological model. These values are estimated from physical properties of the watershed adjusted to reproduce system behavior [100]. The hydraulic conductivity controls the total amount of water that will be split into the surface runoff. The hydraulic roughness affects the peak flow and the time to peak and initial saturation is related with the existing humidity into the soil.

18.2   VALIDATION OF HYDROLOGIC MODEL *Vflo*

Scalars are multiplied by these parameter maps to adjust the value in each grid cell while preserving the spatial heterogeneity. The sequence of adjustment was recommended by Vieux and Moreda [100] to minimize the objective function for volume, and then peak flow, obtaining an overall optimal parameter set for the storms. The OPPA procedure for adjustment can be stated as: increasing the volume of the hydrograph is achieved by decreasing hydraulic conductivity, and similar, increasing peak flow is achieved by decreasing hydraulic roughness. Several adjustments were made when it was necessary to produce consistent results at the USGS stations compared with every storm.

The reference hydrographs were developed from point observations or observed data of USGS stations numbers: #50144000 at Rio Grande de Añasco (San Sebastián), #50136400 at Rio Rosario (Hormigueros) and #50138000 at Rio Guanajibo (Hormigueros) (U.S Geological Survey – Current Water Data for Puerto Rico [93]) and compared with results from the hydrological model.

The ground surface optimum resolution in the model was 200 meters. This was based on the previous studies by Prieto [61] and Rojas [69].

The watershed parameters were adjusted upstream of the observed point (USGS flow stations) by the adjustment method described by Vieux and Moreda [100]. They employ a scalar to adjust parameter maps so that the proposal scalar magnitudes change while the spatial variation is preserved. The scalar used to multiply the n, K and θ parameter maps area is defined as follows [29]:

$$N_{ii} = \frac{1}{8}(2 + 3i)\Big|_{ii=0.1.2.3.4} \tag{1}$$

N_{ij} is the adjustment factor, where the n, K and θ values can be perturbated from 25% to 175%. Study model sensitivity was done for the watershed to identify response sensitivity for peak flow to each storm changing the multiplicative factor in the parameters. The events evaluated were the same 10 events presented in Table 18.6. A list of parameter ensembles is created for each storm in every station (Figure 18.1). A total of 450 simulations were done for this analysis.

Figures 18.2–18.4 present spider plots of rate of change for peak flow using five different adjustment factors in the roughness parameter. The three USGS stations were taken to perform this analysis if the given station recorded the corresponding event. It is possible to observe that when the roughness factor decreases, the rate of change increases and show a higher change. When the adjustment factor is >1, the range of change in peak flow falls and tends to remain constant or with a minimum change in the peak flow, just below the referenced value.

Similarly, results for effect of the hydraulic conductivity are presented in Figure 18.5. Here the maximum rate of change takes to place for the minimum values of hydraulic conductivity. These results are consistent with statements presented in *Gourley and Vieux* [29].

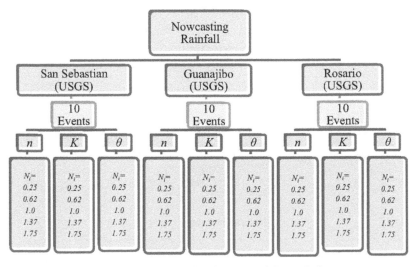

FIGURE 18.1 Flow chart of the calibration factor panel for peak flow.

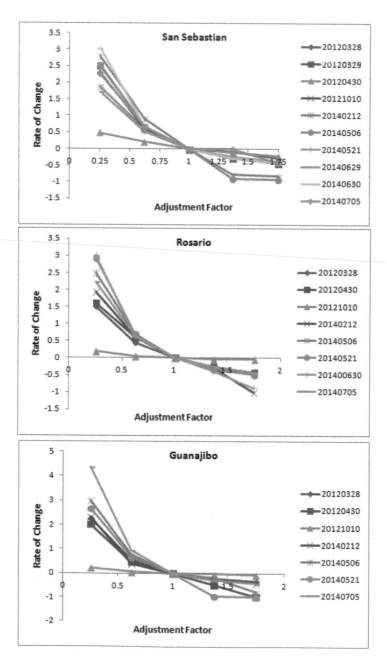

FIGURE 18.2   Spider plot for rate of change of peak flow changing the adjustment factor in the roughness parameter. Dates of rainfall events are shown in the legend.

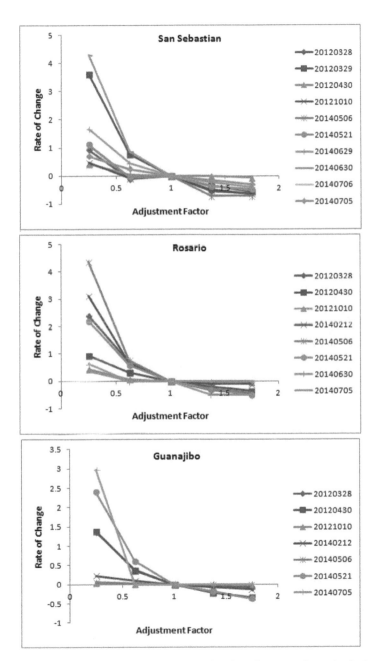

FIGURE 18.3   Spider plot for peak flow changing the adjustment factor in the hydraulic conductivity parameter. Dates of rainfall events are shown in the legend.

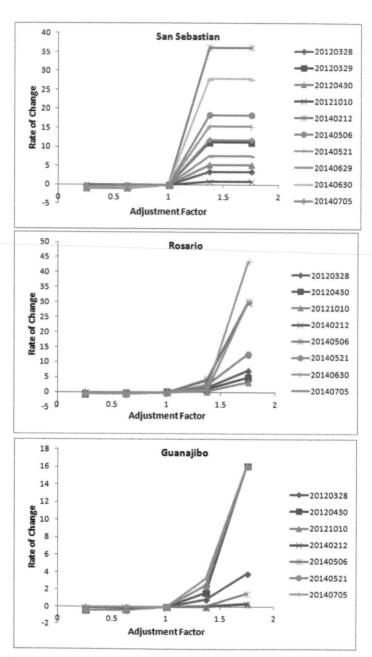

FIGURE 18.4   Spider plot of peak flow changing the adjustment factor in the initial saturation parameter. Dates of rainfall events are shown in the legend.

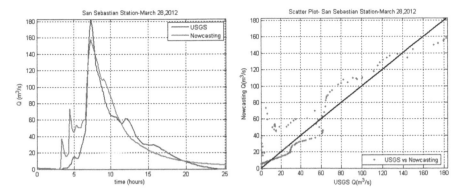

FIGURE 18.5   The left panel is runoff observed data (USGS) blue line and simulated data (Nowcasting) red line at San Sebastián station on March 28, 2012. The right panel is a scatter plot of USGS vs. Nowcasting.

Figure 18.6 shows the results of the sensitivity analysis for the initial saturation. In this case one can observe that for adjustment factor <one, the peak flow presents few changes or continues constant. When the factor adjustment in the initial saturation is 1.37, the peak flow increases excessively and becomes independent of the initial saturation for higher values. The hydrological model with the current characteristics is most sensitive to initial saturation parameters. It may be due to the more presence of clay in the watershed. The clay is included in the soil group D. This group has soils with high potential runoff and very low infiltration capacity, when they are saturated.

The analysis suggests that the initial saturation is the parameter with the highest sensitivity in the peak flow for different storms with short duration. Initial saturation is a parameter that depends of how many storms have occurred previously to the studied storm (antecedent soil moisture). Different results are possible to obtain with a sample of continuous storms.

Similar results were found in peak flow with variations of roughness and hydraulic conductivity for all events. Low variations were found in peak flow when the adjustment factor takes values greater than one.

A compilation of individual simulations is determined based on comparison with the observed stream flow data from (U.S. Geological Survey-Current Water Data for Puerto Rico [93]). The hydrologic evaluation consists of making multiples runs, setting the sensitive parameters in each

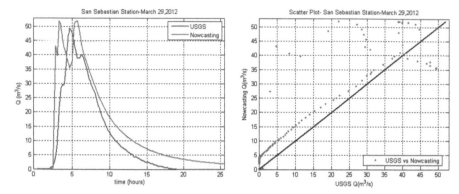

FIGURE 18.6 The left panel is runoff observed data (USGS) blue line and simulated data (Nowcasting) red line at San Sebastián station on March 29, 2012. The right panel is a scatter plot of USGS vs. Nowcasting.

event, yielding the best simulation between observed data from USGS and estimated data from the nowcasting model. The matching of both peaks in every storm was successfully accomplished with flow values.

The separation base flow method used in the USGS stations was the straight line method. It is achieved by joining with a straight line the beginning of the surface runoff to a point on the recession limb representing the end of the direct runoff. Comparison results indicate that the nowcasting model is capable of estimating hydrographs at distributed positions within a watershed based on knowledge of hydrographs at USGS stations. The hydrograph shape is observed with high accuracy, with rising and falling limbs, and hydrograph peaks timed well. Small adjustment between 0.8 and 1.20 were present in the calibration factor. Figure 18.7 presents the hydrograph (left panel) of observed data from the San Sebastián USGS station compared with the simulated results using the nowcasting approach in the hydrological model *Vflo*.

Figures 18.6–18.16 show that the USGS hydrograph at San Sebastián station compared well with the nowcasting hydrograph for the events recorded. The right panel shows a scatter plot of the relation USGS vs. nowcasting results, for different events.

The Mean Square Error (MSE) and Root Mean Square Error (RMSE) analyzes were performed in order to directly determine the effectiveness of the connection between the hydrological model and the rainfall nowcasting model for various events and durations. Results for the MSE showed varying degrees of both overestimation and underestimation for

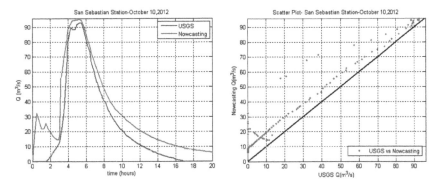

FIGURE 18.7   The left panel is runoff observed data (USGS) blue line and simulated data (Nowcasting) red line at San Sebastián station on October 10, 2012. The right panel is a scatter plot of USGS vs. Nowcasting.

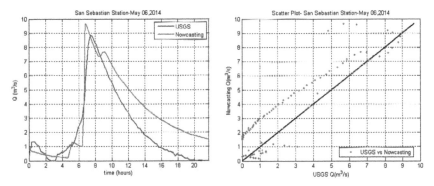

FIGURE 18.8   The left panel is runoff observed data (USGS) blue line and simulated data (Nowcasting) red line at San Sebastián station on May 06, 2014. The right panel is a scatter plot of USGS vs. Nowcasting.

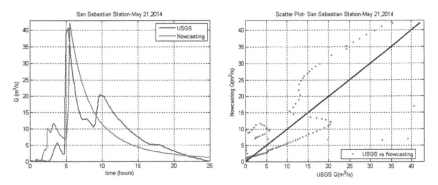

FIGURE 18.9   The left panel is runoff observed data (USGS) blue line and simulated data (Nowcasting) red line at San Sebastián station on May 21, 2014. The right panel is a scatter plot of USGS vs. Nowcasting.

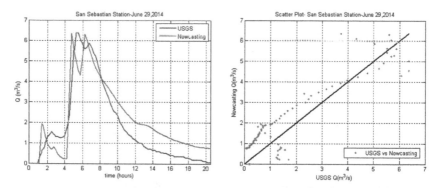

FIGURE 18.10   The left panel is runoff observed data (USGS) blue line and simulated data (Nowcasting) red line at San Sebastián station on June 29, 2014. The right panel is a scatter plot of USGS vs. Nowcasting.

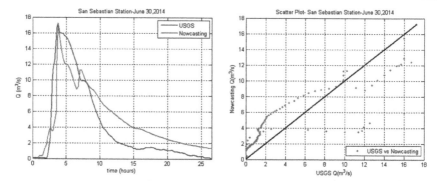

FIGURE 18.11   The left panel is runoff observed data (USGS) blue line and simulated data (Nowcasting) red line at San Sebastián station on June 30, 2014. The right panel is a scatter plot of USGS vs. Nowcasting.

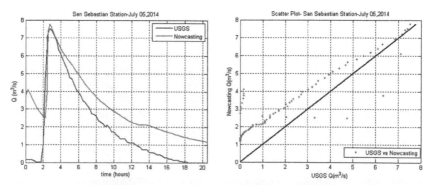

FIGURE 18.12   The left panel is runoff observed data (USGS) blue line and simulated data (Nowcasting) red line at San Sebastián station on July 05, 2014. The right panel is a scatter plot of USGS vs. Nowcasting.

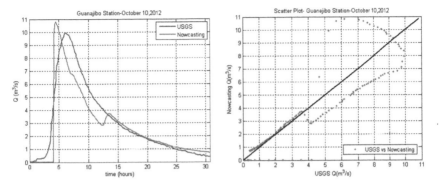

FIGURE 18.13 The left panel is runoff observed data (USGS) blue line and simulated data (Nowcasting) red line at Guanajibo station on October 10, 2012. The right panel is a scatter plot of USGS vs. Nowcasting.

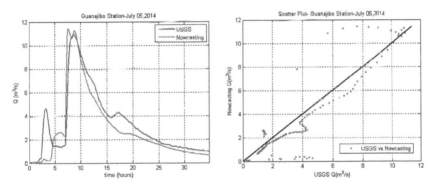

FIGURE 18.14 The left panel is runoff observed data (USGS) blue line and simulated data (Nowcasting) red line at Guanajibo station on July 05, 2014. The right panel is a scatter plot of USGS vs. Nowcasting.

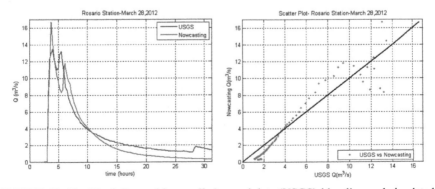

FIGURE 18.15 The left panel is runoff observed data (USGS) blue line and simulated data (Nowcasting) red line at Rosario station on March 28, 2012. The right panel is a scatter plot of USGS vs. Nowcasting.

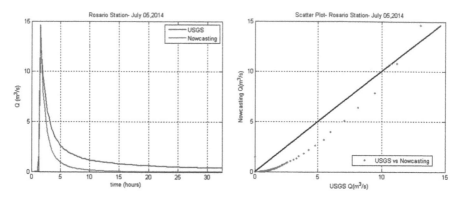

FIGURE 18.16   The left panel is runoff observed data (USGS) blue line and simulated data (Nowcasting) red line at Rosario station on July 05, 2014. The right panel is a scatter plot of USGS vs. Nowcasting.

the various storm events in the three different basins: San Sebastián, Guanajibo and Rosario.

Table 18.1 presents the statistic results at San Sebastián stations. The analysis compares the runoff between the hydrological model using the rainfall forecast and the observed data provided by USGS. Only eight events were considered for this analysis, because the USGS observed data were not available for the events on April 30, 2012 and February 12, 2014. Table 18.2 presents the hydrological statistic results between comparisons: observed data and estimated data for the two events analyzed in the Guanajibo station.

Figures 18.15 and 18.16 present the comparison between data and results on March 28, 2012 and July 05, 2014, respectively at Rosario station USGS. Table 18.3 shows the statistical results using the hydrological model at Rosario station USGS. It is important to note that the most rainfall occurred into the area nearest to Rio Añasco, comprising the San Sebastián station.

The event on July 05, 2014 had superior results than the previous events with respect to the tendency, peak flow and runoff at Rosario station. Results indicate that global nowcasting model can be used to estimate the shape, timing and magnitude of hydrographs.

TABLE 18.1 Hydrological Statistic Results at San Sebastián Station

USGS - Nowcasting	MSE (m³/s)²	RMSE (m³/s)
March 28, 2012	330.7064	18.1853
March 29, 2012	85.9784	9.2724
October 10, 2012	97.1203	9.8549
May 06, 2014	2.3539	1.5342
May 21, 2014	54.3021	7.3690
June 29, 2014	0.7535	0.8680
June 30, 2014	7.8139	2.7953
July 05, 2014	1.3781	1.1739

TABLE 18.2 Hydrological Statistic Results at Guanajibo Station

USGS – Nowcasting	MSE (m³/s)²	RMSE (m³/s)
October 10, 2012	1.8137	1.3467
July 05, 2014	2.1434	1.4640

TABLE 18.3 Hydrological Statistic Results at Rosario station

USGS – Nowcasting	MSE (m³/s)²	RMSE (m³/s)
March 28, 2012	2.9655	1.7220
July 05, 2014	0.9432	0.9711

FLOOD ALERT SYSTEM USING HIGH-RESOLUTION RADAR RAINFALL DATA: INUNDATION (FLOOD) ANALYSIS[1, 2]

LUZ E. TORRES MOLINA

CONTENTS

19.1   INTRODUCTION

The probabilistic flood forecast developed in this research together with the inundation model is capable of providing a forecast of when and where river banks are likely to be overtopped. This could be more detailed with several cross sections into the river.

Decisions for evacuation can be categorized by determining the risk that overtopping represent to residents in areas adjacent to rivers or stream flows. The available knowledge when the evacuation decision can be made include probabilistic flood forecast published by each zone or location

---

[1] This chapter is an edited version from: "*Luz E. Torres Molina, 2014. Flood Alert System Using Rainfall Data in the Mayagüez Bay Drainage Basin, Western Puerto Rico. PhD Thesis, Department of Civil Engineering and Surveying, University of Puerto Rico, Mayagüez Campus*".

[2] Numbers in brackets refer to the references at the end of this book.

with large historical floods. Furthermore, it is then associated with the relevant topographical and demographical information for the basin and river, and the cost associated with the flooding and evacuation.

The approach of Flood Alert System (FAS) is to minimize loss of life and disruptions to communities through identification of the evacuation decision and strategy that has the maximum expected value under current conditions. The potential cost related with the decision model for evacuation can be categorized as losses resulting from preventable flood damage and losses from evacuation. The first is associated with deaths and injuries. Potential damage to building and property should not be considered when making an evacuation decision, as this damage is the same regardless of whether an evacuation is ordered or not. Losses from evacuation refer to evacuation and emergency services, cost associated with the inconvenience, and that associated with the vacating of houses and buildings. Using a FAS model and an adequate flooding history, it is possible to determine a potential evacuation savings or amount of money saved as a result of no evacuation.

19.2   INUNDATION (FLOOD) ANALYSIS

Inundation Analysis is a *Vflo* extension that provides images and animation showing the extent of forecast inundation, which can be used an indication of flood risk [103].

To show the full potential of this tool in enhancing the visualization of the flood area, the program was run with a large storm data. Figure 19.1 presents a time-series flow for the basin area on March 28, 2012. The area north was the most affected by the rainfall on this event. Inundation Analysis presents an inundation sequence each hour. Other events were modeled using inundation animation, but the March 28, 2012 event is good enough to show the potential of this tool. The flow depths results from *Vflo* model were introduced into the inundation to create the animation flow. The animation flow is attached as a link in: http://www.mediafire.com/download/l42s3nbpprk08ib/Appendix.zip.

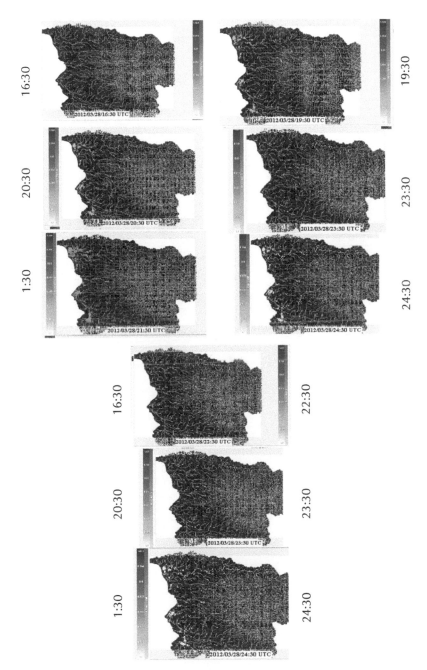

FIGURE 19.1 Inundation sequence each hour, on March 28, 2012 (from the left to right and top to bottom).

CHAPTER 20

FLOOD ALERT SYSTEM USING HIGH-RESOLUTION RADAR RAINFALL DATA: CONCLUDING REMARKS[1,2]

LUZ E. TORRES MOLINA

CONTENTS

20.1 INTRODUCTION

This chapter concludes the research study presented in Part II. It indicates that TropiNet radar technology has been used first time for hydrologic analyzes and specifically for rainfall forecasting in Puerto Rico. Results

---

[1] This chapter is an edited version from: "*Luz E. Torres Molina, 2014. Flood Alert System Using Rainfall Data in the Mayagüez Bay Drainage Basin, Western Puerto Rico. PhD Thesis, Department of Civil Engineering and Surveying, University of Puerto Rico, Mayagüez Campus*".

[2] Numbers in brackets refer to the references at the end of this book.

from the nowcasting model at spatial and temporal scales demonstrated the capability of the model to reproduce observed rainfall, for each nowcasting lead-time with relatively good agreement.

The best statistical results were found in the rainfall nowcasting model with a lead-time of 10 min. It is well known that prediction of sudden storms using rainfall nowcasting models represent the category that are the most difficult to predict, and consequently, providing accurate flash flood warnings from these types of storms is a major challenge.

The nowcasting model has a limitation in the time shift, because it is assumed that the cloud is a rigid object and that the cloud speed is constant, when in reality these parameters can vary. To find the actual weather conditions, more atmospheric parameters should be taken into account. In fact, cloud speed depends on its formation, and other physical parameters that are constantly changing [11]. These factors should be taken into account in future works.

Several parameter estimations were developed at each spatial and temporal domain, and the stochastic behavior of rainfall intensity was represented by an exponential time and spatial lag model, which is an approximation of a stochastic transfer function.

The rainfall nowcasting algorithm searches for contiguous rain pixels and identifies rain cells in the last two radar images to estimate the cloud motion vector. This newly developed rainfall nowcasting algorithm was validated with ten storms and results comparing the algorithm with observed data as well as the hydrological results showed that the nowcasting model is a suitable tool for predicting the most likely areas to become inundated.

Comparisons between rain gauges, TropiNet and NEXRAD demonstrated that the TropiNet radar system provides a higher degree of accuracy in rainfall estimation compared to NEXRAD. The *RMSE* was increased for heavy rain conditions, nevertheless in all cases (light, moderate and heavy rain), TropiNet consistently yielded the smallest error compared with rain gauges, while NEXRAD produced the largest errors. This was the first attempt to evaluate a rainfall prediction in the western Puerto Rico area. The most hydrological sensitive parameter in the basin area is the initial saturation.

When the hydrologic model was evaluated within the Mayagüez bay drainage basin with three USGS reference stations, the San Sebastian station showed the highest flow. The events under analysis presented more rainfall in the north basin area.

Use of a GOES based satellite remote sensing product allowed for the spatial and temporal distribution of potential evapotranspiration input (PET, mm/h) in the hydrologic model. The post-processing algorithm developed in this study provided the ability to change the *PET* size resolution through interpolation.

Differences in the order of 0.75 and 330 *MSE* percent between the observed data from USGS and the results of hydrological model may be due to initial conditions prior to storms, such as soil moisture and daily evapotranspiration distribution.

A study of flood levels can be conducted with the model in the future within the study watershed to estimate flood depths resulting from embankment overtopping, thereby providing recommendations for improving current flood hazard maps.

The nowcasting model was evaluated with the available events from TropiNet radar, but it was also developed to work with events with high precipitation. At the same order, the hydrological model was evaluated in this study with relatively small flow (180 m³/s), but can be evaluated with extraordinary events when they occur. Unfortunately, during the study period, there were no high precipitation events. The data for this research is available in the link below: http://www.mediafire.com/download/l42s3nbpprk08ib/Appendix.zip.

20.2 STUDY LIMITATIONS

The nowcasting model presents a time shift limitation in the prediction of 10, 20 and 30 min. This can be a linear trend in the given data. In the future, an algorithm may be configured to fix the time shift using more than 10 events within the study area.

Figure 20.1 presents a prototype of result with the time shift correction in the average rainfall for all pixels for a time interval of 10 min. In this

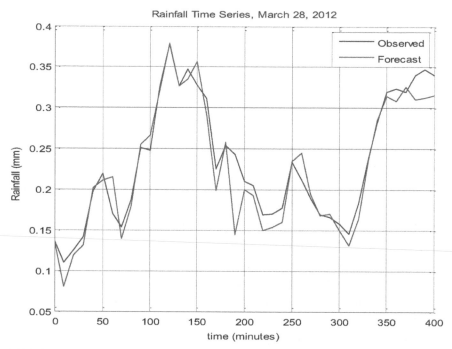

FIGURE 20.1 Adjusted average rainfall plots using a bias correction factor in the time variable on March 28, 2012.

case, Figure 7.7 in Chapter 17 would look like Figure 20.1. The Figure 20.2 presents a prototype of results with the time shift correction in the accumulated rainfall average for all rain pixels during the complete event on March 28, 2012. It can be seen that the bias in accumulated rain is reduced with this time shift correction. The events selected were limited to the data available in the TropiNet radar server. It is recommended to extend both methodologies to high precipitation events.

20.3   FUTURE WORK

The nowcasting model and hydrological model can be evaluated with extreme event data. When the three TropiNet radars are finally operating as a network, they will provide higher resolution data that can be used in the nowcasting model and hydrological model. Using a bias correction in

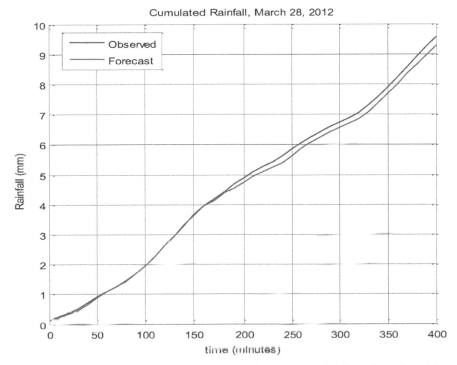

FIGURE 20.2 Adjusted accumulated precipitation average rainfall results using a bias correction factor in the time on March 28, 2012.

the time shift, it is possible to make predictions more accurately. To implement this, an extensive number of events, physical and atmospheric data will be necessary.

20.4   SUMMARY OF RESEARCH STUDY IN PART II

Floods are one of the most costly types of natural disasters in the world. The current work is an attempt to introduce a *Flood Alert System* in the western Puerto Rico, using radars with high temporal and spatial resolution and developing a forecasting model for convective precipitation for time periods of a few hours or less (*nowcasting*).

The accuracy of these forecasts generally decreases very rapidly during the first 30 min because of the very short lifetime of individual convective pixels. A number of observational studies have shown that individual convective cells have mean lifetime of about 20 min, with best performance associated with a lead-time of 10 min. Numerical simulation studies have contributed significantly to the understanding of storm composition and duration; this is just beginning to be recognized in current nowcasting systems. In Part II, a review of the literature is provided related to what is currently known from numerical and observational studies about the organization, lifetime and motion of storms.

The nowcasting technique proposed in this chapter is a special kind of nonlinear model with stochastic and deterministic components. The rainfall forecasts obtained using the considered method is then routed through a rainfall runoff model *Vflo*. Thus, a coupled rainfall-runoff forecasting procedure can be implemented for a watershed in western Puerto Rico. The prediction results with lead-time of 10, 20 and 30 min were analyzed and compared using statistical methods. The forecast result with lead-time of 10 min is the best alternative with least percent of error. It was used in the hydrological model *Vflo* to compare the estimated hydrograph with the observed hydrograph from USGS stations. Furthermore, it was used in the flooding model *Inundation Animator* to show the extent of flooding superimposed onto a land map.

ACKNOWLEDGEMENTS

The author would like to thank her parents for their unconditional support and love. She wishes to express her deep gratitude to her advisor Dr. Sandra Cruz-Pol, who was a great motivator until the end of the project. She always enjoyed academic discussions as well as nonacademic matters. Thanks to Dr. Sandra Cruz-Pol and Dr. Jose Colom for their financial support from two important projects "CASA" and "MRI". Material support was also received from the NOAA CREST Project (NOAA/EPP Grant # NA11SEC4810004). She would like to thank Dr. Eric W. Harmsen for his willingness to take her as his Ph.D. student during the preliminary years of her Ph.D. She would also like to express her sincere appreciation to Dr. Nazario D. Ramirez-Beltran, who was her guide on the nowcasting part for this research. Certainly, a large part of the project was only possible thanks to his contribution, guidance and supervision.

KEYWORDS IN PART II

- **CASA radar**
- **deterministic components**
- **hydrograph**
- **hydrological model**
- **Inundation Animator**
- **land map**
- **Puerto Rico**
- **rainfall runoff model**
- **rainfall runoff model** *Vflo*
- **rainfall-runoff forecasting**
- **statistical model**
- **stochastic**
- **stochastic components**
- **time series analysis**
- **tropical hydrology**
- **USGS**
- **water resources**
- **watershed**
- **WSR-88D**

REFERENCES

PART I

1. Abbott, M. B., Bathurst, J. C., Cunge, J. A., O'Connell, P. E., & Rasmussen, J. (1986a). An introduction to European hydrological system: Systeme Hydrologique Europeen, (SHE), Part I – History and philosophy of a physically based distributed modeling system. *J. Hydrol., 87,* 45–59.
2. Abbott, M. B., Bathurst, J. C., Cunge, J. A., O'Connell, P. E., & Rasmussen, J. (1986b). An Introduction to the European Hydrological System—Systeme Hydrologique European, 'SHE'. 2. Structure of a physically based distributed modeling system. *J. Hydrol., 87,* 61–77.
3. Anderson, T. W., & Darling, D. A. (1954). A Test of Goodness-of-Fit. *Journal of the American Statistical Association, 49,* 765–769.
4. Allen, R. G., Pereira, L. S., Raes, D., & Smith, M. (1998). Crop Evapotranspiration Guidelines for Computing Crop Water Requirements. FAO Irrigation and Drainage Paper 56, Food and Agriculture Organization of the United Nations, Rome.
5. Atger, F. (1999). The skill of ensemble predictions systems. *Monthly Weather Review, 127,* 1941–1953.
6. Baeck, M. L., & Smith, J. A. (1998). Estimation of heavy rainfall by the WSR-88D. *Weather Forecasting, 13,* 416–436.
7. Ball, J. E., & Luk, K. C. (1998). Modeling the spatial variability of rainfall over a catchment. ASCE, *Journal of Hydrologic Engineering, 3*(2), 122–130.
8. Battan, L. J. (1973). *Radar Observation of the Atmosphere.* University of Chicago Press, 323 pp.
9. Bear, J. (1972). *Dynamics of Fluids in Porous Materials.* American Elsevier, 784 pp.
10. Bedient, P. B., Hoblit, B. C., Gladwell, D. C., & Vieux, B. E. (2000). NEXRAD Radar for Flood Prediction in Houston. *Journal of Hydrologic Engineering, 5*(3), 269–277.
11. Bedient, P. B., Holder, A., Benavides, J. A., & Vieux, B. E. (2004). Radar-Based Flood Warning System Applied to Tropical Storm Allison. *Journal of Hydrologic Engineering, 8*(6), 308–318.
12. Bell, V. A., & Moore, R. J. (2000). The sensitivity of catchment runoff models to rainfall data at different spatial scales. *Hydrology and Earth System Sciences, 4*(4), 653–667.
13. Bevan, K. J., & Hornberger, G. M. (1982). Assessing the effect of spatial pattern of precipitation in modeling streamflow hydrographs. *Water Resources Bulletin, 18*(5), 823–829.
14. Beven, K., & Binley, A. (1992). The future of distributed models: Model calibration and uncertainty prediction. *Hydrological Process, 6,* 279–298.

15. Binley, A. M., & Beven, K. J. (1991). Physically based modeling of catchment hydrology: a likelihood approach to reducing predictive uncertainty. In: Farmer, D. G., Rycroft, M. J. (Eds.). *Computer Modelling in the Environmental Sciences*. Clarendon Press, Oxford, pp. 75–88.

16. Birikundavyi, S., & Rouselle, J. (1997). Use of partial duration series for single-station and regional analysis of floods. *Journal of Hydrologic Engineering, 2*, 68–75

17. Bloschl, G., Reszler, C., & Komma, J. (2008). A spatially distributed flash flood forecasting model. *Environmental Modelling & Software, 23*(4), 464–478.

18. Bormann, H. (2006). Impact of spatial data resolution on simulated catchment water balances and model performance of the multiscale TOPLATS model. *Hydrology and Earth System Sciences, 10*(2), 165–179.

19. Bouwer, H. (1966). Rapid field measurement of air entry value and hydraulic conductivity of soil as significant parameters in flow system analysis. *Water Resources Research 2*(4), 729–738.

20. Brasington, K., & Richards, K. (1998). Interactions between model predictions, parameters and DTM scales for TOPMODEL. *Comput. Geosci., 24*(4), 299–314.

21. Burnash, R. J. C., Ferral, R. L., & McGuire, R. A. (1973). A generalized stream flow simulation system—Conceptual modeling for digital computers. Tech. Rep. to the Joint Federal and State River Forecast Center, U.S. National Weather Service and California Department of Water Resources, Sacramento, 204 pp.

22. CASA (2006). CASA Annual Report Year *3*, Volume II. Engineering Research Center for Collaborative Adaptive Sensing of the Atmosphere. Cooperative Agreement No. EEC-0313747, National Science Foundation. March 20.

23. Carpenter, T. M., & Georgakakos, K. P. (2004). Continuous streamflow simulation with the HCDHM distributed hydrologic model. *Journal of Hydrology, 298*, 61–79.

24. Casale, R., & Margottini, C. (2004). Natural Disasters and Sustainable Development. Springer, 398 pp.

25. CRIM, (1998). Center for Municipal Tax Revenues of Puerto Rico. *Digital Elevation Model*.

26. Cole, S. J., & Moore, R. J. (2009). Distributed hydrological modeling using weather radar in gauged and ungauged basins. *Advances in Water Resources, 32*(7), 1107–1120.

27. Droegemeier, K. K., & Smith, J. D. (2000). Hydrological aspects of weather prediction and flood warning: Report of the Ninth Prospectus Development Team of the U.S. Weather Research Program. *Bull. Amer. Meteor. Soc., 81*, 2665–2680.

28. Entekahbi, D., Anderson, M. P., Avissar, R., Bales, R., Hornberger, G. M., Nuttle, W. K., Parlange, M. B., Peters-Lidard, C., Potter, K. W., Roads, J. O., Wilson, J. L., & Wood, E. F. (2002). Report of a Workshop on Predictability and Limits to prediction in Hydrologic Systems. Committee on Hydrologic Science, National Research Council, National Academy Press, ISBN 0–309–08347–8. pp. 118.

29. Epstein, E. S. (1969). A scoring system for probability forecast of ranked categories. *Journal of Applied Meteorology, 8*, 985–987.

30. Fang, Z., & Bedient, P. B. (2007). NEXRAD Flood Warning System and Floodplain Library for Houston, TX. ASCE Conf. Proc.: *World Environmental and*

Water Resources Congress: Restoring our Natural Habitat. doi:http://dx.doi.org/10.1061/40927(243)302

31. Fishman, G. S. (1996). Monte Carlo: Concepts, Algorithms and Applications. Springer, New York, USA. 723 pp.

32. FEMA, (2009). Flood Insurance Study. Commonwealth of Puerto Rico, Volume 1 of *5,* Revised: Nov. *18,* 2009. Federal Emergency Management Agency

33. Freer, J., Beven, K., & Ambroise, B. (1996). Bayesian estimation of uncertainty in runoff prediction and the value of data: An application of the GLUE approach. *Water Resour. Res., 32,* 2161–2173.

34. Freeze, R. A., & Cherry, J. A. (1979). Groundwater. Prentice Hall Publishers, 604 pp.

35. Figueroa-Alamo, C., Aquino, Z., Guzmán-Ríos, S., & Sánchez, A. V. (2006). *Water resources data Puerto Rico and the U.S. Virgin Islands.* Water Year 2004: U.S. Geological Survey Water-Data Report PR-04-1, 597 p.

36. Georgakakos, K. P. (2006a). Analytical results for operational flash flood guidance. *Journal of Hydrology, 317,* 81–103.

37. Georgakakos, K. P. (2006b). Hydrologic Short Term Forecasting with QPF Input. White *Paper in Proceedings of USWRP Warm Season Precipitation Workshop,* 5–7 March 2002, National Center for Atmospheric Research, Boulder, Colorado, 5 pp.

38. Gourley, J. J., & Vieux, B. E. (2005). A method for evaluating the accuracy of quantitative precipitation estimates from a hydrologic modeling perspective. *American Meteorological Society.* April, 115–133.

39. Gourley, J. J., & Vieux, B. E. (2006). A method for identifying sources of model uncertainty in rainfall-runoff simulations. *Journal of Hydrology, 327,* 68–80.

40. Goyal, M. R., González, E. A., & C. Chao de Báez, (1988). Temperature versus elevation relationships for Puerto Rico. *J. Agric. UPR 72*(3), 449–67.

41. Griensven, van, A., Meixner, T., Grunwald, S., Bishop, T., Diluzio, M., & Srinivasan, R. (2006). A global sensitivity analysis tool for the parameters of multivariable catchment models. *Journal of Hydrology 324,* 10–23.

42. Gritmit, E. P., & Mass, C. F. (2002). Initial results of a mesoscale short-range ensemble forecasting system over the Pacific North-west. *Weather and Forecasting, 17,* 192–205.

43. Gupta, R. S. (1989). Hydrology and Hydraulic Systems. Waveland Press, Inc., Illinois. 302–307 pp. ISBN 0-88133-865-6.

44. Hamill, T. M., Whitaker, J. S., & Wei, X. (2004). Ensemble reforecasting: improving medium-range forecast skill using a retrospective forecast. *Monthly Weather Review, 132,* 1434–1447.

45. Hargreaves, G. H., & Samani, Z. A. (1985). Reference crop evapotranspiration from temperature. *Appl. Eng. Agric.,* ASAE. *1*(2), 96–99.

46. Harmsen, E. W., Converse, J. C., & Anderson, M. P. (1991). Application of the Monte-Carlo Simulation Procedure to Estimate Water-Supply Well/Septic Tank-Drainfield Separation Distances in the Central Wisconsin Sand Plain. *Journal of Contaminant Hydrology, 8*(1), 91–109.

47. Harmsen, E. W., Gomez Mesa, S. E., Cabassa, E., N. D. Ramirez Beltran, Cruz-Pol, S., Kuligowski, R. J., & Vasquez, R. (2008). Satellite subpixel rainfall variability. *WSEAS Transactions on Signal Processing, 8*(7), 504–513.

48. Harmsen, Mecikalski, E. W. J., Cardona-Soto, M. J., A. Rojas González, & Vásquez, R. (2009). Estimating daily evapotranspiration in Puerto Rico using satellite remote sensing. *WSEAS Transactions on Environment and Development*, *6*(5), 456–465.

49. Harmsen, E. W., Mecikalski, J., Mercado, A., & Tosado Cruz, P. (2010). Estimating evapotranspiration in the Caribbean Region using satellite remote sensing. *Proceedings of the AWRA Summer Specialty Conference, Tropical Hydrology and Sustainable Water Resources in a Changing Climate*. San Juan, Puerto Rico. August 30–September 1.

50. HCFCD, (1984). *Criteria Manual for the Design of Flood Control and Drainage Facilities*. Houston, Texas, Harris County Flood Control District.

51. Helmer, E. H., Ramos, O., López, T. M., Quiñónez, M., & Díaz, W. (2002). Mapping the forest type and land cover of Puerto Rico, a component of the Caribbean biodiversity hotspot. *Caribbean Journal of Science*, *38*(3/4), 165–183.

52. Henderson, F. M. (1966). *Open Channel Flow*. Macmillan, New York. 544 pp.

53. Hydrologic Engineering Center, (2006). HEC-HMS, Hydrologic Modeling System. U.S. Army Corps of Eng, Davis, CA.

54. Hydrologic Engineering Center, (2008). HEC-RAS River Analysis System. Version 4.0 U.S. Army Corps of Eng, Davis, CA.

55. Julien, P. Y., & Saghafian, B. (1991). CASC2D user manual—a two dimensional watershed rainfall–runoff model. Civil Engineering Report CER90–91PYJ-BS-12, Colorado State University, Fort Collins, p. 66.

56. Julien, P. Y., Saghafian, B., & Ogden, F. L. (1995). Raster-based hydrological modeling of spatially varied surface runoff. *Water resources bulletin*. AWRA *31*(3), 523–536.

57. Kondragunta, C. R., Kitzmiller, D., Seo, D. J., & Shestha, K. (2005). Objective integration of satellite, rain gauge, and radar precipitation estimates in the Multisensor Precipitation Estimator algorithm. *19th Conference on Hydrology, Amer. Meteor Soc.,* Seattle.

58. Kondragunta, C. R., & Shestha, K. (2006). Automated real-time operational rain gauge quality control tools in NWS hydrologic operations. *20th Conference on Hydrology, 86th Amer. Meteor Soc. Annual Meeting,* Atlanta, GA.

59. Leavesley, G. H., Lichty, R. W., B. M. Troutman., & Saindon, L. G. (1983). Precipitation-runoff modeling system user manual. USGS Water Resources Investigations Rep. No. 83–4238, Denver.

60. Lawrence, B. A., Shebsovich, M. I., Glaudemans, M. J., & Tilles, P. S. (2003). Enhancing precipitation estimation capabilities at National Weather Service field offices using multisensor precipitation data mosaics. *83rd AMS Annual Meeting, 19th International Conference on Interactive Information Processing Systems for Meteorology, Oceanography, and Hydrology*, Long Beach, California, February 9–13.

61. Maddox, R. A., Hoxit, L. R., Chappell, C. H., & Caracena, F. (1978). Comparison of the meteorological aspects of the Big Thompson and Rapid City flash floods. *Mon. Wea. Rev., 106,* 375–389.

62. Mahani, S. E., & Khanbilvardi, R. (2009). Generating Multi-Sensor Precipitation Estimates over radar gap areas. *WSEAS Transactions on Systems, 8,* 96–106.

63. Marshall, J. S., & Palmer, W. (1948). The distribution of raindrops with size. *J. Meteor., 5,* 165–166

64. McGregor, K. C., Bingner, R. L., Bowie, A. J., & Foster, G. R. (1995). Erosivity index values for northern Mississippi. *Trans. ASCE, 38,* 1039–1047.

65. McWhorter, D. B., & Sunada, D. K. (1977). Ground-Water Hydrology and Hydraulics. Water Resources Publications. 304 pp.

66. McMichael, C. E., Hope, A. S., & Loaiciga, H. A. (2005). Distributed hydrological modeling in California semiarid shublands: MIKESHE model calibration and uncertainty estimation. *Journal of Hydrology. 317,* 307–324.

67. Minitab Inc. (2010). Meet MINITAB 16 for Windows®. English Version. www.minitab.com

68. Moore, I. D., Grayson, R. B., & Ladson, A. R. (1991). Digital terrain modeling: a review of hydrological, geomorphological and biological applications. *Int. J. Hydrological Process, 5*(1), 3–30.

69. Moore, A. M., & Kleeman, R. (1998). Skill assessment for ENSO using ensemble prediction. *Quarterly Journal of the Royal Meteorological Society, 124,* 557–584.

70. Müller, W. A., Appenzeller, C., Doblas-Reyes, F. J., Liniger, M. A. (2005). A debiased ranked probability skill score to evaluate probabilistic ensemble forecasts with small ensemble sizes. *Journal of Climate 18*(10), 1513–1523

71. Murphy, A. H. (1971). A note on the ranked probabilistic predictions and probability score in the cost-loss ratio situation. *Journal of Applied Meteorology, 5,* 534–537.

72. NOAA, (2006). *Puerto Rico Mean Annual Precipitation 1971–2000.* Hydrology and River Information. National Oceanic and Atmospheric Administration, National Weather Service Forecast Office, San Juan, Puerto Rico, http://www.srh.noaa.gov/sju/?n=mean_annual_precipitation

73. NOAA, (No date). Advanced Hydrologic Prediction Service (Online). Available: http://water.weather.gov/ahps2/hydrograph.php?wfo=sju&gage=horp4

74. Ogden, F. L., & Julien, P. Y. (1994). Runoff model sensitivity to radar rainfall resolution. *J. Hydrol., 158,* 1–18.

75. Prieto, M. G. (2006). Development of a Regional Integrated Hydrologic Model for a Tropical Watershed. Master of Science Thesis, University of Puerto Rico at Mayagüez, PR.

76. PRWRERI, (2004). Land Use Classification of the Mayagüez Bay Watershed (Río Grande de Añasco, Río Yagüez, and Río Guanajibo Watersheds. Puerto Rico Water Resources and Environmental Research Institute [PRWRERI]. Developed for the Puerto Rico Environmental Quality Board.

77. Quinn, P., Beven, K., Chevallier, P., & Planchon, O. (1991). The prediction of hillslope flow paths for distributed hydrological modeling using digital terrain models. *Hydrol. Processes, 5,* 59–79.

78. Ramírez-Beltran, Kuligowski, N. R. J., Harmsen, E. W., Castro, J. M., Cruz-Pol, S., & Cardona-Soto, M. (2008a). Rainfall estimation from convective storms using the Hydro-Estimator and NEXRAD. *WSEAS Transaction on Systems. 10*(7), 1016–1027.

79. Ramirez-Beltrán, N. D., Kuligowski, R. J., Harmsen, E. W., Castro, J. M., Cruz-Pol, S., & Cardona-Soto, M. J. (2008b). Validation and strategies to improve the Hydro-

Estimator and NEXRAD over Puerto Rico. *12th WSEAS International Conference on SYSTEMS*, Heraklion, Greece, July 22–24.

80. Rawls, W. J., Brakensiek, D. I.., Miller, N. (1983). Green-Ampt Infiltration Parameters from Soils Data. *Journal of Hydraulic Engineering* 109(1), 62 – 70.

81. Reed, S., Koren, V., Smith, M., Zhang, Z., Moreda, F., D. J. Seo and DMIP Participants, (2004). Overall distributed model intercomparison project results. *Journal of Hydrology, 298,* 27–60.

82. Rojas-González, A. M. (2004). Estudio Comparativo sobre hidráulica de ríos en valles aluviales. Master of Science Thesis, University of Puerto Rico at Mayagüez, PR., 140 pp.

83. Rutledge, A. T. (1998). Computer programs for describing the recession of ground-water discharge and for estimating mean ground-water recharge and discharge from streamflow records-update. U.S. Geological Survey Water Resources Investigations Report 98–4148.

84. Sahho, G. B., Ray, C., & De Carlo, E. H. (2006). Calibration and validation of a physically distributed model, MIKE SHE, to predict streamflow at high frequency in a flashy mountainous Hawaii stream. *Journal of Hydrology, 327,* 94–109.

85. Saxton, K. E., & Rawls, W. (2006). Soil Water Characteristics, Hydraulic Properties Calculator. Hydrology and Remote Sensing Laboratory, USDA-ARS, Maryland, USA.

86. Schaap, M. G. (2003). *Rosetta Lite Version 1.1*. George E. Brown Jr. Salinity Laboratory and UC Riverside, Department of Environmental Sciences, June.

87. Schaap, M. G., & Leij, F. J. (1998a). Database related accuracy and uncertainty of pedotransfer functions. *Soil Science 163,* 765–779.

88. Schaap, M. G., & Leij, F. J. (1998b). Using neural networks to predict soil water retention and soil hydraulic conductivity. *Soil & Tillage Research, 47,* 37–42.

89. Schaap, M. G., Leij, F. J., & van Genuchten, M. Th. (1998). Neural network analysis for hierarchical prediction of soil water retention and saturated hydraulic conductivity. *Soil Sci. Soc. Am. J., 62,* 847–855.

90. Schaap, M. G., Leij, F. J., & van Genuchten, M. Th. (2001). Rosetta: a computer program for estimating soil hydraulic parameters with hierarchical pedotransfer functions. *Journal of Hydrology,* 251,163–176.

91. Scofield, R. A., & Kuligowski, R. J. (2003). Status and outlook of operational satellite precipitation algorithms for extreme-precipitation events. *Weather Forecasting, 18,* 1037–1051.

92. Seo, D. J. (1998). Real-time estimation of rainfall fields using rain gauge data under fractional coverage conditions. *Journal of Hydrology, 208,* 25–36.

93. Seo, D. J., Briedenbach, J. P., & Johnson, E. R. (1999). Real-time estimation of mean field bias in radar rainfall data. *Journal of Hydrology, 223,* 131–147.

94. Sepúlveda, N., Pérez-Blair, F., DeLong, L. L., & López-Trujillo, D. (1996). *Real-Time Rainfall-Runoff Model of the Carraízo-Reservoir Basin in Puerto Rico*, Water-Resources Investigations Report 95–4235, U.S. Geological Survey. San Juan, Puerto Rico.

95. Shestha, R. K., Yasuto, T., & Kaoru, T. (2002). IC ratio concept in distributed hydrological modeling for optimal performance. *International Conference on Urban Hydrology for the twenty-first century,* 14–16 October, Kuala Lumpur, 790–800.

96. Simunek, J., M.Th. van Genuchten, & Sejna, M. (2005). *Code for Simulating the one-dimensional movement of water, heat, and multiple solutes in variably saturated porous media,* Department of Environmental Sciences, University of California Riverside and US Salinity Laboratory, USDA, ARS, Riverside, CA, USA.

97. Smith, M. B., Seo, D. J., Koren, V. I., Reed, S., Zhang, Z., Duan, Q. Y., Moreda, F., & Cong, S. (2004). The distributed model intercomparison project (DMIP): motivation and experiment design. *J. Hydrol. DMIP Special Issue.*

98. Spear, R. C., & Hornberger, G. M. (1980). Eutrophication in Peel Inlet II. Identification of critical uncertainties via generalized sensitivity analysis. *Cybernetics, 14,* 43–49.

99. Stensrud, D. J., Bao, J. W., & Warner, T. T. (2000). Using ensembles for short range forecasting. *Monthly Weather Review, 127,* 433–446.

100. Stephens, M. A. (1970). Use of the Kolmogorov-Smirnov, Cramer-Von Mises and related statistics without extensive tables. *Journal of the Royal Statistical Society. Series B (Methodological), 32*(1), 115–122.

101. Sweeney, T. L. (1992). *Modernized Areal Flash Flood Guidance.* NOAA Technical Report NWS HYDRO 44. Hydrologic Research Laboratory, National Weather Service, NOAA, Silver Spring, MD, October, 21 pp. and Appendix I.

102. Tarboton, D. G., Bras, R. L., & Rodriguez-Iturbe, I. (1991). On extraction of channel networks from digital elevation data. *Hydrolog. Process, 5,* 81–100.

103. U.S. Census Bureau, (2010). Census 2010 Data for Puerto Rico. U.S. Census Bureau.

104. U.S. Army Corps of Engineers, (1990). HEC-1 Flood Hydrograph Package. Hydrologic Engineering Center, Davis, California, September.

105. USACE, (1996). Hydrologic Aspects of Flood Warning Preparedness Programs. Department of the Army, U.S. Army Corps of Engineers, Engineering Technical Letter, ETL 1110-2-540.

106. U.S. Department of Agriculture, Natural Resources Conservation Service, (2006a). Soil Survey Geographic (SSURGO) database for Mayagüez area. Puerto Rico Western Part, pr684, Fort Worth, Texas, Publication date: Dec. 26. http://SoilDataMart.nrcs.usda.gov/

107. U.S. Department of Agriculture, Natural Resources Conservation Service, (2006b). Soil Survey Geographic (SSURGO) database for Lajas Valley Area. Puerto Rico. Fort Worth, Texas, Publication date: Dec. 26. http://SoilDataMart.nrcs.usda.gov/

108. U.S. Department of Agriculture, Natural Resources Conservation Service, (2006c). Soil Survey Geographic (SSURGO) database for Arecibo area. Puerto Rico Western Part, PR682, Fort Worth, Texas, Publication date: Dec. 26. http://SoilDataMart.nrcs.usda.gov/

109. U.S. Department of Agriculture, Natural Resources Conservation Service, (2006d). Soil Survey Geographic (SSURGO) database for Ponce area. Puerto Rico Western Part, PR688, Fort Worth, Texas, Publication date: Dec. 26. http://SoilDataMart.nrcs.usda.gov/

110. U.S. Department of Commerce, (1961). Technical Paper *42,* Generalized Estimates of Probable Maximum Precipitation and Rainfall-Frequency Data for Puerto Rico and Virgin Islands. National Oceanic and Atmospheric Administration, Weather Bureau.

111. USGS, (1999). Estimation of Magnitude and Frequency of Floods for Streams in Puerto Rico: New Empirical Models. Water Resources Investigation Report 99–4142, U.S. Department of the Interior, Geological Survey, San Juan, Puerto Rico.

112. Lopez, M. A. (1977). U.S. Department of the Interior, Geological Survey. Regional Flood Frequency for Puerto Rico. Open-File Report.

113. Van der Perk, M., & Bierkens, M. F. P. (1997). The identifiability of parameters in a water quality model of the Biebzra river, Poland. *Journal of Hydrology 200*, 307–322.

114. van Genuchten, M.Th. (1980). A closed-form equation for predicting the hydraulic conductivity of unsaturated soils. *Soil Sci. Am. J. 44*, 892–898.

115. Viessman, W., & Lewis, G. L. (1996). *Introduction to Hydrology*. 4th Edition, Harper-Collins, 760 pp.

116. Vieux, B. E. (1988). Finite Element Analysis of Hydrologic Response Areas Using Geographic Information Systems. PhD Thesis, Department of Agricultural Engineering, Michigan State University, July, 199 pp.

117. Vieux, B. E., Bralts, V. F., Segerlind, & Wallace, R. B. (1990). Finite element watershed modeling: one-dimensional elements. *J. Water Resour. Plan. Mgmt.*, 116(6), 803–819.

118. Vieux, B. E. (1993). Surface runoff modeling. *Computing, 7*(3), 310–338.

119. Vieux, B. E., & Gauer, N. (1994). Finite element modeling of storm water runoff using GRASS GIS. *Microcomput.* Civil Eng. 9(4), 263–270.

120. Vieux, B. E., & Farajalla, N. S. (1996). Temporal and Spatial Aggregation of NEXRAD Rainfall Estimates on Distributed Hydrologic Modeling. *Third Int. Conf. on GIS and Environmental Modeling*, Santa Fe, NM, Nat. Center Geog. Info. Annual, 199– 204.

121. Vieux, B. E., & Bedient, P. B. (1998). Estimation of rainfall for flood prediction from WSR-88D reflectivity: A case study, 17–18 October 1994. *American Meteorological Society*. June, 407–415.

122. Vieux, B. E. (2001). Distributed Hydrologic Modeling Using GIS. Water Science Technology Series, Kluwer Academic Publishers, Norwell, Massachusetts, 311 pp. ISBN 0-7923-7002-3.

123. Vieux, B. E., Ester, C., Dempsey, C., & Vieux, J. E. (2002). Prospects for a real-time flood warning system in Arizona. *Proceedings of the 26th Annual Conference, Breaking the Cycle of Repetitive Flood Loss, Association of State Floodplain Managers*, June 23–28, 2002, Phoenix, AZ.

124. Vieux, B. E., & Vieux, J. E. (2002). Vflo™: A real-time distributed hydrologic model. *Proceedings of the Second Federal Interagency Hydrologic Modeling Conference*, July 28–August 1, Las Vegas, NV 2002. Abstract and paper in CD-ROM.

125. Vieux, B. E., Chen, C., Vieux, J. E., & Howard, K. W. (2003). Operational Deployment of a Physics-based Distributed Rainfall-runoff Model for Flood Forecasting in Taiwan. *IAHS General Assembly at Sapporo*, Japan, July 3–11, 2003.

126. Vieux, B. E., & Moreda, F. G. (2003). Ordered physics-based parameter adjustment of a distributed model. In: Duan, Q., Sorooshian, S., Grupta, H. V., Rousseau, A. N., Turcotte, R. (Eds.), *Water Science and Application Series*, vol. 6. American Geophysical Union, pp. 267–281, ISBN 0-87590-355-X (Chapter 20).

127. Vieux, B. E., & Associates, Inc. (2004). *VfloTM 3.0, Desktop User Manual*. www.vieuxinc.com.

128. Vieux, B. E., Cui, Z., & Gaur, A. (2004). Evaluation of a physics-based distributed hydrologic model for flood forecasting. *Journal of Hydrology*, *298*, 155–177.

129. Vieux, B. E., & Bedient, P. B. (2004). Assessing urban hydrologic prediction accuracy though event reconstruction. *Journal of Hydrology*, *299*, 217–236.

130. Vieux, B. E., Bedient, P. B., & Mazroi, E. (2005). Real-time urban runoff simulation using radar rainfall and physics-based distributed modeling for site-specific forecasts. *10th International Conference on Urban Drainage*, Copenhagen, Denmark, 21–26 August.

131. Vieux, B. E., & Vieux, J. E. (2005). Statistical evaluation of a radar rainfall system for sewer system management.

132. Vieux, B. E., & Imgarten, J. M. (2006). On the scale-dependent propagation of hydrologic uncertainty using high-resolution X-band radar rainfall estimates. *Atmospheric Research*, *103*, 96–105.

133. Vieux, B. E., & Vieux, J. E. (2006). Evaluation of a physics-based distributed hydrologic model for coastal, island, and inland hydrologic modeling. In: *Coastal Hydrology and Processes*. Water Resources Publications, LLC, Highlands Ranch, CO, USA, pp. 453–464.

134. Viglione, A., Chirico, G. B., Komma, J., Woods, R., Borga, M., & Blöschl, G. (2010). Quantifying space-time dynamics of flood event types. *Journal of Hydrology*, *394*(1–2), 213–229. Elsevier B. V. doi:10.1016/j.jhydrol.2010.05.041

135. Villalta-Calderón, C. A. (2004). Selección de Funciones de Transporte de Sedimentos para los Ríos de la Bahía de Mayagüez usando SAM. Master of Science Thesis, University of Puerto Rico at Mayagüez, PR., 441 pp.

136. Warner, T. T., Brandes, E. A., Sun, J., Yates, D. N., & Mueller, C. K. (2000). Prediction of a flash flood in complex terrain. Part I: A comparison of rainfall estimates from radar, and very short range rainfall simulations from a dynamic model and an automated algorithmic system. *J. Appl. Meteor.*, *39*, 797–814.

137. Wechsler, S. (2006). Uncertainties associated with digital elevation models for hydrologic applications : a review. *Hydrol. Earth Syst. Discuss.*, *3*, 2343–2384.

138. Wunderground web page, (No Date). WenderMap. Avialable: http://www.wunderground.com/US/PR/ [2010, June 5].

139. Western, A., Zhou, S., Grayson, R., Mcmahon, T., Bloschl, G., & Wilson, D. (2004). Spatial correlation of soil moisture in small catchments and its relationship to dominant spatial hydrological processes. *Journal of Hydrology*, *286*(1–4), 113–134. doi: 10.1016/j.jhydrol.2003.09.014.

140. Whiteaker, T., Robayo, O., Maidment, D., & Obenour, D. (2006). From a NexRAD rainfall map to a flood inundation map. *Journal of Hydrologic Engineering*, *11*(1), 37–45.

141. Wigmosta, M. S., Vail, L. W., & Lettenmaier, D. P. (1994). A distributed hydrology-vegetation model for complex terrain. *Water Resour. Res. 30*(6), 1665–1679.

142. Wilks, D. S. (1995). Statistical Methods in the Atmospheric Sciences: An Introduction. Academic Press, San Diego, 467 pp.

143. Wilson, J. W., & Brandes, E. A. (1979). Radar measurement of rainfall A summary. *Bulletin of the AMS*, *60*(9), 1048–1058.

144. Woodley, W. L., Olsen, A. R., Herndon, A., & Wiggert, V. (1975). Comparison of gauge and radar methods of convective rain measurement. *Journal of Applied Meteorology*, *14*, 909–928.

PART II

1. Akaike, H. (1974). A new look at the statistical model identification. *IEEE Transactions on Automatic Control, 19,* 716–723.
2. Anagnostou, E. N., & Krajewski, W. F. (1998). Calibration of the WSR-88D precipitation processing subsystem. *Weather and Forecasting, 13,* 396–406.
3. Arocho Meaux, S., Mercado-Vargas, A., Pablos-Vega, G. A., Harmsen, E. W., Cruz-Pol, S., & J. Colom Ustáriz, (2010). Calibration and validation of CASA radar rainfall estimation. *Proceedings of the AWRA Summer Specialty Conference, Tropical Hydrology and Sustainable Water Resources in a Changing Climate.* San Juan, Puerto Rico. August 30 to September 1.
4. Boggs, P. T., & Tolle, J. W. (1996). Sequential quadratic programming. *Acta Numerica, 4,* 1–51.
5. Box, G., & Jenkins, G. (1976). Time Series Analysis. In: *Forecasting and Control Review Ed. Holden-Day.,* p. 575.
6. Breidenbach, J. P., Seo, D. J., & Fulton, R. A. (1998). Stage II and III post processing of the NEXRAD precipitation estimates in the modernized National Weather Service. *14th Conf on IIPS,* AMS, Phoenix.
7. Browning, K. A., & Collier, C. G. (1989). Nowcasting of precipitation systems. *Reviews of Geophysics, 27*(3), 345–370.
8. Burlando, P., Rosso, R., Cadavid, L., & Salas, J. D. (1993). Forecasting of Short-Term Rainfall using ARMA Models. *J. Hydrol., 144,* 193–211.
9. Burnash, R. J. C., Ferral, R. L., & McGuire, R. A. (1973). *A generalized streamflow simulation system—Conceptual modeling for digital computers.* Tech. Rep. to the Joint Federal and State River Forecast Center, *U.S. National Weather Service and California Department of Water Resources,* Sacramento, 204 pp.
10. Chow, V. T., Maidment, D., & Mays, L. W. (1988). *Applied Hydrology.* McGraw Hill Book Co.
11. Corfidi, S. F., Merritt, J. H., & Fritsch, J. M. (1996). Predicting the movement of mesoscale convective complexes. *Weather Forecasting,* 11, 41–46.
12. Cruz-Pol, S., Colom, J., Córdoba, M., Pablos, G., Ortiz, J., Castellanos, W., Acosta, M., & Trabal, J. (2011). Climatic radars developed at uUniversity of Puerto Rico – Mayaguez Campus (*Red de Radares Meteorológicos Desarrollados en el RUM*). Magazine Dimension by CIAPR (*Revista Dimensión Ingeniería y Agrimensura by CIAPR*), *25*(1), 13–17.
13. Davis, J. C. (1986). *Statistics and Data Analysis in Geology.* John Wiley and Sons, New York.
14. De Luca, D. L. (2005). Metodi di previsione dei campi di pioggia. Tesi di Dottorato di Ricerca, Universit'a della Calabria, Italy.
15. Delleur, J. W., & Kavvas, M. L. (1978). Stochastic Models for Monthly Rainfall Forecasting and Synthetic Generation. J. *Appl. Meteor., 17,* 1528–1536.
16. Delleur, J. W., Chang, T. J., & Kavvas, M. L. (1989). Simulation models of sequences of wet and dry days. *J. Irrig. Drain. Eng.* ASCE, *115,* 344–357.
17. DHI, 2005. *MIKE SHE Technical Reference.* DHI Water and Environment, Danish Hydraulic Institute (DHI), Denmark.

18. Diaz Gonzalez, M. F. (2012). *Estimation of Water Balance and Groundwater Processes of the Salinas to Patillas Area in South-eastern Puerto Rico for 1980–2010*. Master Thesis. Department of Civil Engineering, University of Puerto Rico – Mayagüez Campus. Pp. 177.

19. Dizon, C. Q. (2010). ARMA modeling of a stochastic process appropriate for the Angat reservoir. *Philippine Engineering Journal, 48*(1),June.

20. Dixon, M., & Weiner, G. (1993). TITAN: Thunderstorm identification, tracking, analysis, and nowcasting: a radar-based methodology. *Journal of Atmospheric and Ocean Technology, 10*(6), 785–797.

21. Einfalt, T., Denoeux, T., & Jacquet, G. (1990). A radar rainfall forecasting method designed for hydrological purposes. *Journal of Hydrology, 114*, 229–244.

22. FAO, (2014). Penman -Monteith. Food and Agriculture Organization of the United Nations. http://www.fao.org/docrep/x0490e/x0490e08.htm

23. FEMA, (2012). *Flood insurance study*. Commonwealth of Puerto Rico, Volume 1 of 5, Preliminary: June 22. Federal Emergency Management Agency.

24. French, M. N., Krajewski, W. F., & Cuykendall, R. R. (1992). Rainfall forecasting in space and time using a neural network. *J. Hydrol., 137*, 1–31.

25. Galvez, M. B., Colom, J. G., Chandrasekar, V., Junyent, F., Cruz-Pol, S., Rodríguez-Solís, R. A., León, L., Rosario-Colon, J. J., De Jesús, B., Ortiz, J. A., & Mora Navarro, K. M. (2013). First Observations of the Initial Radar Node in the Puerto Rico Tropinet X-Band Polarimetric Doppler Weather Testbed. *IGARSS*, 2337–2340.

26. Gautier, C., Diak, G. R., & Masse, S. (1980). A simple physical model to estimate incident solar radiation at the surface from GOES satellite data. *J. Appl. Meteor., 19*, 1007–1012.

27. Georgakakos, K. P. (2006). Analytical results for operational flash flood guidance. *Journal of Hydrology, 317*, 81–103.

28. Google Earth, http://www.google.com/earth/index.html.

29. Gourley, J. J., & Vieux, B. E. (2005). A method for evaluating the accuracy of quantitative precipitation estimates from a hydrologic modeling perspective. *American Meteorological Society*, April, 115–133.

30. Goyal, M. R., González, E. A., & Chao de Báez, C. (1988). Temperature versus elevation relationships for Puerto Rico. *J. Agric. UPR, 72*(3), 449–467.

31. Gupta, R. S. (1989). *Hydrology and Hydraulic Systems*. Waveland Press, Inc., Illinois, 302–307 pp.

32. Haire, W. J. (1972). Flood in the Rio Guanajibo Valley, South-western Puerto Rico. U.S. Geological Survey Hydrologic Investigation Atlas HA – 456.

33. Hargreaves, G. H. (1975). Moisture availability and crop production. *Transactions of the ASAE, 18*(5), 980–984.

34. Hargreaves, G. H., & Samani, Z. A. (1982). Estimating potential evapotranspiration. *Journal of Irrigation and Drainage Division*, Proceedings of the ASCE, *108*(IR3), 223–230.

35. Harmsen, E. W., Mecikalski, J., Cardona-Soto, M. J., Rojas González, A., & Vásquez, R. (2009). Estimating daily evapotranspiration in Puerto Rico using satellite remote sensing. *WSEAS Transactions on Environment and Development, 6*(5), 456–465.

36. Harmsen, E. W., Mecikalski, J., Mercado, A., & Tosado Cruz, P. (2010). Estimating evapotranspiration in the Caribbean Region using satellite remote sensing. *Proceed-*

ings of the AWRA Summer Specialty Conference, Tropical Hydrology and Sustainable Water Resources in a Changing Climate. San Juan, Puerto Rico. August 30– September 1.

37. Hoppe, R. H. W., Linsenmann, C., & Petrova, S. I. (2006). Primal-dual Newton methods in structural optimization. *Comp. Visual. Sci., 9,* 71–87.

38. Hyndman, R. J. (2010). Forecasting overview. *International Encyclopedia of Statistical Science,* 536–539.

39. Johnson, R. C., J. C. Imboff., Kittle, J. L., & A. S. Donigian Jr. (1984). *Hydrological Simulation Program-Fortran, User's Manual for Release 8.0.* U.S. Environmental Protection Agency, Athens, GA.

40. Kashyap, R. L., & Rao, A. R. (1976). *Dynamic Stochastical Models from Empirical Data.* Academic Press.

41. Katz, R. W., Skaggs, R. H. (1981). On the Use of Autoregressive-Moving Average Processes to Model Meteorological Time Series. *Mon. Wea. Rev., 109,* 479–484.

42. Kohnova, S., Svetlíková, D., Komorníková, M., Hlavčová, K., & Szolgay, J. (2007). Analysis of time series of discharges in the region of the Kláštorské Lúky wetland. *Acta Hydrologica Slovaca., 8*(1), 67–78.

43. Koussis, A. D., Lagouvardos, K., Mazi, K., Kotroni, V., Sitzmann, D., Lang, J., Zaiss, H., Buzzi, A., & Malguzzi, P. (2003). Flood forecasts for urban basin with integrated hydro-meteorological model. *J. Hydrologic Eng., 8*(1), 1–11.

44. Leavesly, G. H., & Stannard, L. G. (1995). Chapter 9: The Precipitation Runoff Modeling System – PRSM. In: *Computer Models of Watershed Hydrology* by Singh, V. P., ed., Water Resources Publications. Littleton, CO.

45. Malvic, T., & Balić, D. (2009). Linearity and Lagrange linear multiplicator in the equations of ordinary kriging. *Nafta, 59*(1), 31–37.

46. Marshal, J. S., & Palmer, W. (1948). The distribution of raindrops with size. *J. Meteor., 5,* 165–166

47. McLeod, A. I. (1993). Parsimony, model adequacy and periodic correlation in forecasting time series. *International Statistical Review, 61,* 387–393.

48. Merriam-Webster's online dictionary (2012).

49. MathWorks, Inc. (2011). *Optimization Toolbox for use With MatLab.* The MathWorks Inc., Mayagüez, Puerto Rico.

50. Mujumdar, P. P., & Nagesh, D. (1990). Stochastic models of streamflow: some case studies. *Hydrological Sciences Journal, 35*(4), 395–410.

51. NDFD, (2010). *National Weather Service National Digital Forecast Database.* (http://www.weather.gov/forecasts/graphical/sectors/puertorico.php).

52. NCDC, (2013). National Climatic Data Center – National Oceanic and Atmospheric Administration (NOAA). (http://www.ncdc.noaa.gov/).

53. NOAA-NWS, (1995). South-east Texas tropical mid-latitude rainfall and flood event. Natural Disaster Survey Report, 50 pp. [Available from U.S. Department of Commerce, NWS/NOAA, *National Weather Service,* Southern Region Headquarters, 819 Taylor Street, Room 10A26, Fort Worth, TX 76102.]

54. Novo, S. (2008). Immediate prediction of convective storms by radar: Update, (*Pronostico inmediato de tormentas convectivas por radar: una actualización*). *Rev. bras. meteorol., 23*(1), 41–50.

55. NRCS, (2014). Natural resource Conservation Service, Water and Climate Center http://www.wcc.nrcs.usda.gov/scan/.
56. Personal Communication, *Alejandra Rojas*, University of Puerto Rico at Mayagüez (2012).
57. Personal Communication, *Ernesto Rodriguez*, NWS, San Juan (2010).
58. Personal Communication, *Carlos Ansemi,* NWS, San Juan (2013, 2014).
59. Personal Communication, Prof. Eric Harmsen, University of Puerto Rico at Mayagüez (2014).
60. Pielke, R. A. (1984), *Mesoscale Meteorological Modeling*. New York, NY: Academic Press, 612 pp.
61. Prieto, M. G. (2007). *Development of a Regional Integrated Hydrologic Model for a Tropical Watershed.* Master of Science Thesis, University of Puerto Rico at Mayagüez, PR.
62. Ramirez-Beltran, N. D. (1996). A vector autoregressive model to predict hurricane tracks. *International Journal of Systems Science, 27*(1), 1–10.
63. Ramírez-Beltran, D, N., Castro, J. M., Harmsen, & Vasquez, R. F. (2008). Stochastic transfer function models and neural networks to estimate soil moisture. *Journal of the American Water Resources Association, 44*(4), 847–865.
64. Ramirez-Beltran, N. D., Castro, J. M., & Gonzalez, J. (2014). An algorithm for predicting the spatial and temporal distribution of rainfall rate. *International Journal of Water*.
65. Ramirez, J., & Bras, R. L. (1985). Conditional distributions of Neyman-Scott models for storm arrivals and their use in irrigation control. *Water Resour. Res., 21*(3), 317–330.
66. Reklaitis, G. V., Ravindran, A., & Ragsdell, K. M. (1983). *Engineering Optimization: Methods and Applications*. John Wiley and Sons.
67. Rinehart, R. E. (1997). *Radar for Meteorologists*. 3rd Ed, 1997, Rinehart Publications.
68. Robinson, J. S., Sivapalan, M., & Snell, J. D. (1995). On the relative roles of hillslope processes, channel routing, and network geomorphology in the hydrological response of natural catchments, *Water Resour. Res., 31*, 3089–3101.
69. Rojas, A. M. (2012). Flood *Prediction Limitations in Small Watersheds with Mountainous Terrain and High Rainfall Variability*. Doctor of Philosophy in Civil Engineering, University of Puerto Rico at Mayagüez, PR.
70. Rodriguez-Iturbe, I., Cox, D. R., & Isham, V. (1987). Some models for rainfall based on stochastic point process. *Proc. R. Soc. London*, Ser. A., *410*, 269–288.
71. Rossa, A., Bruen, M., Frühwald, D., Macpherson, B., Holleman, I., Michelson, D., & Michaelides, S. (2005). *Use of Radar Observations in Hydrological and NWP Models*. COST 717 Final Report.
72. Rossa, A., Laudanna Del Guerra, F., Borga, M., Zanon, F., Settin, T., & Leuenberger, D. (2010). Radar-driven High-resolution hydrometeorological forecasts of the 26 September 2007 Venice flash flood. *Journal of Hydrology*.
73. Ryzhkov, A. V., & Zernic, D. S. (1995). Comparison of dual-polarization radar estimators of rain. *Journal of Atmospheric and Oceanic Technology, 12*, 249–256.
74. Salas, J. D., Delleur, J. W., Yevjevich, V., & Lane, W. L. (1980). *Applied Modeling of Hydrologic Time Series*. Water Resources Publications, Littleton, Colorado, p.484.

75. Salas, J. D., & Obeysekera, J. (1992). Conceptual basis of seasonal streamflow time series models. *ASCE J. Hydraulic Eng.*, *118*(8), 1186–1194.
76. Schell, G. S., Madramootoo, C. A., Austin, G. L., Broughton, R. S. (1992). Use of radar measured rainfall for hydrologic modeling. *Canadian Agricultural Engineering*, *34*(1), 41–48.
77. Seo, D. J., & Breidenbach, J. P. (2002). Real-time correction of spatially nonuniform bias in radar rainfall data using rain gauge measurements, *Journal of Hydrometeorology*, *3*, 93–111.
78. SERCC, (2013). South-east Regional Climate Center. University of North Carolina, Chapel Hill, NC, http://www.sercc.com/cgi-bin/sercc/cliMAIN.pl?pr6073.
79. Sepúlveda, N., Pérez-Blair, F., DeLong, L. L., & López-Trujillo, D. (1996). *Real-Time Rainfall-Runoff Model of the Carraízo-Reservoir Basin in Puerto Rico*. Water-Resources Investigations Report 95-4235, *U.S. Geological Survey*. San Juan, Puerto Rico.
80. Sirangelo, R., Versace, P., & De Luca, D. L. (2007). Rainfall nowcasting by at site stochastic model PRAISE. *Hydrology and Earth System Sciences*, *11*, 1341–1351.
81. Smith, K. T., & Austin, G. L. (2000). Nowcasting precipitation – A proposal for a way forward. *Journal of Hydrology, 239*, 34–45.
82. Schwab, G. O., Fangmeier, D. D., & Elliot, W. J. (1996). *Soil and Water Management Systems*. John Wiley and Sons. New York, NY.
83. Sun, X., Mein, R. G., Keenan, T. D., Elliott, J. F. (2000). Flood estimation using radar and raingauge data. *Journal of Hydrology, 239*, 4–18.
84. Stern, R. D., & Coe, R. (1984). A model fitting analysis of daily rainfall data (with discussion). *Journal of the Royal Statistical Society*, Series A, *147*, 1–34.
85. Trabal, J. M., Pablos-Vega, G. A., Ortiz, J. A., Colom-Ustáriz, J. G., Cruz-Pol, S., McLaughlin, D. J., Zink, M., & Chandrasekhar, V. (2011). Off-the-Grid weather radar network for precipitation monitoring in western Puerto Rico. *Proceedings of the International Symposium on Weather Radar and Hydrology*, Exeter, UK, April.
86. Toth, E., Brath, A., & Montanari, A. (2000). Comparison of short-term rainfall prediction models for real-time flood forecasting, *Journal of Hydrology, 239*, 132–147.
87. Ulbrich, C. W., & Lee, L. G. (1999). Rainfall measurement error by WSR-88D radars due to variations in Z-R law parameters and the radar constant. *Journal of Atmospheric and Oceanic Technology, 16*, 1017–1024.
88. U.S. Census Bureau, (2010). Census 2010 Data for Puerto Rico. U.S. Census Bureau.
89. USDA, U.S. Department of Agriculture, Natural Resources Conservation Service, (2006). *Soil Survey Geographic (SSURGO) database for Mayagüez area*. Puerto Rico Western Part, pr684, Fort Worth, Texas, Publication date: Dec. 26. http://SoilDataMart.nrcs.usda.gov/.
90. USDA, U.S. Department of Agriculture, Natural Resources Conservation Service, (2006). *Soil Survey Geographic (SSURGO) database for Lajas Valley Area*. Puerto Rico Western Part, pr687 Fort Worth, Texas, Publication date: Dec. 26. http://SoilDataMart.nrcs.usda.gov/.
91. USDA, U.S. Department of Agriculture, Natural Resources Conservation Service, (2006). *Soil Survey Geographic (SSURGO) database for Arecibo area*. Puerto. Rico Western Part, pr682, Fort Worth, Texas, Publication date: Dec. 26. http://SoilDataMart.nrcs.usda.gov/.

92. USDA, U.S. Department of Agriculture, Natural Resources Conservation Service, (2006). *Soil Survey Geographic (SSURGO) database for Ponce area.* Puerto Rico Western Part, pr688, Fort Worth, Texas, Publication date: Dec. 26. http://SoilDataMart.nrcs.usda.gov/.

93. U.S Geological Survey, (2014). Current Water Data for Puerto Rico. http://waterdata.usgs.gov/pr/nwis/rt.

94. USGS, (2010). Real Time Flood Alert System (RTFAS) for Puerto Rico. Fact Sheet 2010–3029, May.

95. Viessman, W., & Lewis, G. L. (1996). *Introduction to Hydrology.* 4th Edition, Harper-Collins, 760 pp.

96. Vieux, B. E., & Bedient, P. B. (1998). Estimation of rainfall for flood prediction from WSR-88D reflectivity: A case study, 17–18 October 1994. *American Meteorological Society,* June, 407–415.

97. Vieux, B. E. (2001). *Distributed Hydrologic Modeling Using GIS. Water Science Technology Series.* Kluwer Academic Publishers, Norwell, Massachusetts, 311 pp.

98. Vieux, B. E., & Vieux, J. E. (2002). *Vflo™: A real-time distributed hydrologic model. Proceedings of the Second Federal Interagency Hydrologic Modeling Conference,* July 28–August 1, Las Vegas, NV.

99. Vieux, B. E., Ester, C., Dempsey, C., & Vieux, J. E. (2002). Prospects for a real-time flood warning system in Arizona. *Proceedings of the 26th Annual Conference, Breaking the Cycle of Repetitive Flood Loss, Association of State Floodplain Managers,* June 23–28, Phoenix, AZ.

100. Vieux, B. E., & Moreda, F. G. (2003). Ordered physics-based parameter adjustment of a distributed model. In: Duan, Q., Sorooshian, S., Grupta, H. V., Rousseau, A. N., Turcotte, R. (Eds.), *Water Science and Application Series,* volume 6. American Geophysical Union, pp. 267–281 (Chapter 20).

101. Vieux, B. E., & Vieux, J. E. (2006). Evaluation of a physics-based distributed hydrologic model for coastal, island, and inland hydrologic modeling, In *Coastal Hydrology and Processes. Water Resources Publications,* LLC, Highlands Ranch, CO, USA, pp. 453–464.

102. Vieux, B. E. (2013). Hydrology and Radar Rainfall. http://www.vieuxinc.com.

103. Villalta, C. A. (2004). Selection of transport functions for transport of sediments : The rivers of Mayaguez Bay using SAM (*Selección de Funciones de Transporte de Sedimentos para los Ríos de la Bahía de Mayagüez usando SAM*). Master of Science Thesis, University of Puerto Rico at Mayagüez, PR., 441 pp.

104. Wilks, D. S. (1995). *Statistical Methods in the Atmospheric Sciences: An Introduction.* Academic Press, San Diego, 467 pp.

105. Wilson, J. W., & Brandes, E. A. (1979). Radar measurement of rainfall: A summary. *Bulletin of the AMS, 60*(9), 1048–1058.

106. Xin, Y., & Gang, X. (2009). *Linear Regression Analysis: Theory and Computing.* Hackensack, NJ, USA.

107. Xplorah Project, (2010). School of Planification (*Escuela de Planificación*). University of Puerto Rico – Rio Piedras Campus.

108. Yilmaz, K., Hogue, T. S., Hsu, K., Sorooshian, S., Gupta, H. V., & Wagener, T. (2005). Inter-comparison of rain gauge, radar and satellite-based precipitation esti-

mates with emphasis on hydrologic forecasting. *Journal of Hydrometeorology*, *6*(4), 497–517

109. Yunhao, C., Xiaobing, L., & Peijun, S. (2001). Estimation of regional evapotranspiration over North-west China by using remotely sensed data. *Journal of Geophysical Sciences*, *11*(2), 140–148.

110. Wackernagel, H. (2003). *Multivariate Geostatistics: An Introduction with Applications*. Springer, Third edition, France, pp. 387.

111. Warner, T. T., Brandes, E. A., Sun, J. Z., Yates, D. N., & Mueller, C. K. (2000). Prediction of a flash flood in complex terrain, Part I: a comparison of rainfall estimates from radar, and very short range rainfall simulations from a dynamic model and an automated algorithmic system. *Journal of Applied Meteorology, 39*, 797–814.

112. Westrick, K. J., Mass, C. F., & Colle, B. A. (1999). The Limitations of the WSR-88D Radar Network for Quantitative Precipitation Measurement Over the Costal Western United States. *Bull. Amer. Meteor. Soc.*

113. World Meteorological Organization (WMO), (1981). Flash Flood Forecasting. In: Hall, A. J. (ed.), *Operational Hydrology Report No. 18*, p. 38.

INDEX